Flotsametrics and the Floating World

How One Man's Obsession WITH *Runaway Sneakers* AND *Rubber Ducks Revolutionized Ocean Science*

Smithsonian Books

COLLINS

An Imprint of HarperCollins*Publishers*

Flotsametrics and the Floating World

Curtis Ebbesmeyer and
Eric Scigliano

HarperCollins books may be purchased for educational, business, or sales promotional use. For information, please write: Special Markets Department, HarperCollins Publishers, 10 East 53rd Street, New York, NY 10022.

FIRST EDITION

Designed by Kate Nichols

Library of Congress Cataloging-in-Publication Data

Ebbesmeyer, Curtis C.
Flotsametrics and the Floating World : how one man's obsession with runaway sneakers and rubber ducks revolutionized ocean science / Curtis Ebbesmeyer and Eric Scigliano. — 1st ed.
p. cm.
Includes bibliographical references and index.
ISBN 978-0-06-155841-2
1. Ocean currents. 2. Marine debris. I. Scigliano, Eric, 1953–
II.Title.
GC231'2.E23 2009
551.46'2—dc22

2008038805

09 10 11 12 13 OV/RRD 10 9 8 7 6 5 4 3 2 1

To Susie, who made this journey possible.

Surely the sea
is the most beautiful face in our universe.

— Mary Oliver, "The Waves"

Contents

Preface: A New World

Dave Barry, who often harkens back to his salad days as a reporter covering sewage treatment, isn't the only journalist to recall that unglamorous beat fondly. Call it perverse, but I found the subject fascinating. That was partly because the stakes seemed so high out here: Seattle and the towns around it dump their effluent into Puget Sound, a spectacularly beautiful, fruitful, and fragile body of water. But it was also because I and others covering the subject had Curt Ebbesmeyer to talk to. Ebbesmeyer was the go-to guy on the Sound's mysterious ways, the oceanographer who could explain how its waters moved and how things dumped into them did or didn't get flushed out to the open sea. He was always ready to share what he knew, and to put it in terms that whoever was listening could understand.

Over the years, since we've both moved on from the sewage beat, I've often seen Ebbesmeyer quoted in the local and national papers or heard him answer a radio interviewer's questions. The topics have grown ever stranger: spilled shipping containers, "shoenamis" of sneakers, flotsam flocks of rubber (actually plastic) duckies, drifting corpses, even severed feet. But however exotic these objects might be, they had one quality in common: They all floated on the sea, sometimes for astonishing distances, and in the process revealed oceanic processes as intricate and finely meshed as the workings of a clock or a living organism. Many speak of the sea as a living thing, but for most that's just a metaphor or vague intuition. For Curt Ebbesmeyer it's a concrete reality, to be studied in the same ways a physiologist deciphers the body's processes and a physician diagnoses its ills.

Just as a good doctor learns to read every clue, however unexpected, Ebbesmeyer finds telltale data where others see only trash—in the most literal sense. As he says, every piece of flotsam has a tale to tell—one small piece of the ocean's great story. And anyone who is willing to pay attention, who has the feet and eyes and curiosity to comb a beach, can join in unraveling that story.

I saw how infectious Ebbesmeyer's own curiosity can be—and how much inspirational and instructional effect it can have—at the Beachcombers' Fun Fair held each squally March in the amiably ramshackle resort town of Ocean Shores, Washington. Other scientists might dismiss the participants in such a homespun event as mere hobbyists; beachcombers do pursue their flotsam treasures as avidly as stamp or doll collectors. But Curt sees them as researchers in the rough, potential recruits to a worldwide army of flotsam finders and ocean monitors. The high point of each year's Fun Fair at Ocean Shores is the Dash for Trash, a scavenger hunt-cum-beach cleanup. Scores of trash dashers scatter along the sprawling sands of what tourism boosters bill as the "world's longest ocean drive," fill up heavy black garbage bags, and throng around a folding table where Curt Ebbesmeyer waits.

One by one, they present their booty. He spreads it out and pores over it, patiently explaining each relic's meaning: This plastic tube is an oyster spacer bar, torn loose from a Japanese shellfish farm. This serrated black plastic cone is the cap of a hagfish trap, used to catch eel-like sea-bottom scavengers that are savored in East Asia. These chemical glow sticks are used to attract swordfish and halibut to hooked longlines (as in miles long). Good thing you retrieved this scrap of net before a plus tide dragged it back out to sea, where it might strangle a seal or seabird. Likewise this plastic bag—it only takes one to choke any sea turtle that mistakes it for a jellyfish. And what on earth is *this?*

Ebbesmeyer habitually hunches forward, leaning yet farther in to catch questions from children. His is the posture of a man who's spent much time poring over spread charts and rubble-strewn beaches—or of a bear craning to snuff up a treat. He is a large man, tall and no longer lean, with a shock of white hair that's congenitally tousled and incongruously boyish. His beard covers a receding chin and jowl, and though it's cropped short it lends a Santa Claus quality, an impression abetted by his wide grin and twinkling, bespectacled eyes. He dresses in classic Seattle casual: khaki pants, comfortable shoes, a lightweight parka over loose-fitting plain sweaters, just the threads for a walk on a Northwest beach. He's often mistaken for a prof, though he's worked in the wider world since taking his PhD, feeling more at home on bouncing boats and in the rough-and-tumble of the oil derricks than amid the intrigues of academe. Indeed, academia may be one of the few places Curt Ebbesmeyer does not feel at home.

He listens patiently when others talk; his curiosity ranges far, and people, like flotsam, often have unexpected stories to tell. When he speaks, he emphasizes big points by pausing wide-eyed, brow arched as though shocked at his own impertinence, soliciting assent or disputation before he continues. He exclaims, "Cool!" at good news and interesting ideas and shakes hands in the raised-arm, power-to-the-people clench of the sixties. Somehow, these do not seem like anachronisms or affectations in a sixty-five-year-old man. They're just further signs of his enthusiasm and affability.

Ebbesmeyer's voice is gentle, a gravelly low tenor. It is not a conventional speaker's voice, but he captivates the audiences he addresses, and he addresses many. Though he would blush at such terms, to a far-flung community of beachcombers, ocean watchers, and amateur "flotsamologists," he is a guru and oracle—the man who taught them to read flotsam and love the ocean more deeply. He insists he learns from them.

At the beachcombers' fair, I spoke to a cheerful young man named André Hart and his wife and mother. For them, beachcombing was more than a sideline or diversion; it was a lifeline. In 1993 Hart suffered a severe head injury, courtesy of a drunk driver, and fell into a long coma. For years after he awakened, his mother, Priscilla Hart, explained, "he didn't find anything he could enjoy or get involved in. Then we took him beachcombing. Now he does it all the time. He gets up at four in the morning so he can go scour the beaches. His life just revolves around this, and around *him.*" She nodded toward Ebbesmeyer, who was judging trash. "At first, all we did was look for garbage. Then Dr. Ebbesmeyer enlightened us. We started seeing more there." Now they talk of selling their house, buying a camper, and following the storms and flotsam year-round.

We writers have taken as much in our way from Curt Ebbesmeyer's work as the Harts have. Many have built articles or books on the perambulations of the sneakers and bath toys. Some have appropriated his ideas or incorporated them without knowing where they came from; so familiar has the great "garbage patch" in the North Pacific become that no one seems to remember that it was Curt who coined the term. But what they've had to share is just—the clichés are irresistibly apt here—the tip of the iceberg and the surface of the sea. His most exciting and original work has remained buried in scholarly journals or the files that fill his basement. So widely have Curt's thoughts and research ranged that assembling them in a coherent narrative proved daunting even for him—"like drinking from a fire hose," as he likes to say. It's been a pleasure and a privilege to assist, and to journey with, Curt through the floating world.

Eric Scigliano
Seattle, July 31, 2008

Flotsametrics and the Floating World

I. Chasing Water

I was a penniless, uneducated man.
A piece of driftwood.
—Abraham Lincoln

In the wee hours of May 27, 1990, midway between Seoul and Seattle, the freighter *Hansa Carrier* met a sudden storm and, as freighters often do, lost some of the cargo lashed high atop her deck. Twenty-one steel containers, each forty feet long, tore loose and plunged into the North Pacific. Five of those containers held high-priced Nike sports shoes bound for the basketball courts and city streets of America. One sank to the sea floor. Four broke open, spilling 61,820 shoes into the sea—and into the vast stream of flotsam, containing everything from sex toys to computer monitors, that is released each year by up to ten thousand overturned shipping containers.

One year later, in early June 1991, I stopped by my parents' house in Seattle, as I did every week or so, for lunch and the latest news. My mother, who loved serving as my personal clipping service, had extracted a wire story from the local paper. It reported a strange phenomenon: Hundreds of Nike sneakers, brand-new save for some seaweed and barnacles, were washing up along the Pacific coasts of British Columbia, Washington, and, especially, Oregon, Nike's home state. A lively market had developed; beach dwellers held swap meets to assemble matching pairs of the remarkably wearable shoes, laundered and bleached to remove the sea's traces. The details as to how they'd gotten there were sketchy, verging on nonexistent, and that piqued my mother's curiosity. "Isn't this the sort of thing you study?" she asked, assuming as ever that her son the oceanographer knew everything about the sea. "I'll look into it," I said.

I started looking and never stopped. Seventeen years and many thousands of shoes, bath toys, hockey gloves, human corpses, ancient treasures, and other floating objects later, I'm still looking.

Objects like these have been falling into the sea and washing up on the shores since the dawn of navigation—for billions of years, if you count driftwood, volcanic pumice, and all the other natural materials that float upon the waves. Ordinarily, flotsam is soon lost to human memory—though not, as we shall see, to the ocean's memory. The Great Sneaker Spill would have proved one more curiosity in the annals of beachcombing if my mother hadn't asked her question, and if I hadn't been ready to see the research doors that it opened.

It's only now that I can see how my entire life—from my first childhood encounters with the sea to decades of mainstream research into currents, tides, drifting pollutants, and the curious mobile water bodies called slabs—had prepared me for the puzzle posed by this spill. These thousands of lost sneakers composed a giant scientific experiment on a silver platter, fully if unwittingly funded by Nike—a serendipitous window into the ocean's deepest secrets. They were also the grain around which a worldwide network of beachcombing field volunteers has formed, zealously scouting out and recording telltale washups from Norway to New Zealand.

These high-seas drifters offer a new way of looking at the seas, their movements, and, as we shall see, their music. Call it "flotsametrics." It's led me to a world of beauty, order, and peril I could not have imagined even after decades as a working oceanographer—the floating world.

I did not grow up beside the sea; we lived across the San Rafael Mountains in the hot and dusty San Fernando Valley. My mother and father were

Budding oceanographer Curt Ebbesmeyer, age four, learns about seaweed from his parents at Zuma Beach, California.

raised in Chicago and never saw the ocean until the war brought them to California in 1941. But we were close enough to the water to pine for it—and to escape to the beach whenever we had a free day. Perhaps being so near and yet cut off from the sea made me crave it all the more.

As far back as I can remember, I was fascinated with water and its movements. As soon as I could get my hands on a garden hose, I stuck it in the ground and watched the soil bubble up and wash away around it, like sand on a beach. I would make a pond out of my red Radio Flyer wagon, filling it with water and setting toys and beer bottles floating across it. In elementary school I wrote a story about Paul Bunyan but recast him as a giant of the ocean rather than the woods, striding from sea to sea in his seven-league boots.

My father was a chocolate salesman. Perhaps this followed from his mother's career back in Chicago—making bootleg whiskey, a trade she learned growing up on an Iowa farm and then used to see her children through the Depression after her husband died as the result of an industrial accident. Dad's stock-in-trade was a fine German chocolate brand named Merckens. Twice a month he drove up the coast from Los Angeles to San Francisco teaching small candy shops along the way how to dip conventional American chocolates in melted Merckens. He was a natural at such performances—tall and mirthful, with hair turned a distinguished premature white by all the ether he'd been administered as a teenager during operations on a badly broken ankle. He was a born starter-upper, always organizing projects when he got home—a go-kart for us, new trees for the yard, a block wall around our entire half-acre lot.

Dad's sales trips usually lasted a week, and after each he brought home presents for my brother Scott and me. One Easter, when I was about ten years old, he brought two yellow ducklings. With characteristic whimsy, he named them Flotsam and Jetsam, names that would stay with me for the rest of my life. No one could have guessed how prophetic that gift would prove to be.

Even Dad's chocolate trade seems in retrospect to have forecast the path I would take. The Western world's first chocolate salesman was Christopher Columbus, who brought Europe its first cacao beans when he returned from America. And it was flotsam that led Columbus to America in the first place.

As I grew up, I returned again and again to the water. I took up surfing and scuba, in effect making my body a tool for flotation experiments. The sea was such a presence in my life that I took it for granted; I did not imagine it could be a subject for formal study when the time came to start college.

Choosing a school was no problem; San Fernando Valley State College (now California State University at Northridge) was nearby and cost just $25 a semester. But I had no inkling what I wanted to study there; I felt then, as I've often felt since, like Abraham Lincoln's "piece of driftwood." Mom and Dad couldn't offer much help; they'd never been to college. But I was pretty good at math and science and liked doing projects, so when I visited campus I stopped by the Engineering Department. It was just starting out and eager for good students who could help it win accreditation. The staff persuaded me to take an aptitude test. I got the top score, and they begged me to enroll, which sounded appealing at the time.

I completed my degree in mechanical engineering, but belatedly realized I had no real enthusiasm for it. I did, however, discover one lasting passion at Valley State. In those days, high schools and colleges still required physical education, and I struggled to fit PE credits into my schedule. By my junior year I was working forty hours a week at the phone company while carrying seventeen units of physics, chemistry, and engineering; any PE class I could take would have to start early, and I would have to wear a suit and tie so I could dash off to work afterward. I found a 7:00 AM dance class. Fair enough—my parents had made me take dance classes during high school, and I'd gotten catcalls when I ducked out of track practice early and the coach hollered, "Curt is leaving for his dance class!"

I showed up in my suit for my first college dance class and discovered about three guys and thirty girls had also enrolled. My eyes went straight to one of the women, and I still remember what she wore: white high heels, a pink sleeveless shift and matching sweater—Jackie Kennedy–style—with

It all started at dance class. Young sweethearts Curt and Susie arrive at his fraternity house.

her hair teased into a fashionable beehive. When it came time to select partners, I walked straight over to Susie and asked for the first dance. It was love at first sight, the greatest "Aha!" moment of my life. We danced together through the whole class, and afterward I asked if I could walk her to her next class—quickly, because I had to get to work. From then on we were inseparable. In April 1965, shortly before I graduated, we got married. Soon afterward, Susie introduced me to her great passion, the ballet. In the forty-three years since, we must have attended a hundred performances. I've come to see how the movements of the ocean are a sort of ballet, the drift of flotsam choreographed to the music of the currents.

June 1965 was not an auspicious time to lose a student deferment. Three months earlier the first Marines had gone ashore in Vietnam; by the end of that year 180,000 American soldiers, sailors, and airmen would be serving there. And there was a very good chance I would soon join them.

But that was not the most pressing thought on my mind as I finished college. Before I could worry about whether or not I would go to war, I needed to find a job. And for the first (but far from the last) time, fate in the form of petroleum intervened in my life. It's surprising, and a little shocking, to think how much of my life has been intertwined with oil—the best and worst of substances, which powers ships, produces plastic, fouls the ocean, and enriches and ravages human life like nothing else. Oil has greased the tracks of my life's progress and marked each milestone in my understanding of the floating world.

As graduation approached, I happened to spot a notice on a bulletin board at the Business Department announcing that Standard Oil of New York (later Mobil Oil) would be interviewing on campus. When I showed up, I discovered that only the company's Advertising Department was represented. But the interviewer recommended me to someone who might have something more up my line: Bill Clauser, chief of oil production in Standard/Mobil's Bakersfield District.

Clauser was an old-style Southern gentleman with a well-developed sense of Southern hospitality. My interview with him was an unusual one:

He invited Susie and me over to his Pasadena home and grilled steaks and made margaritas. There's nothing like a margarita to take the chill off an interrogation, and Clauser seemed to like me. I was hired.

In the 1960s, however, engineers at Mobil did not proceed straight to their drafting tables. Everyone who was hired first had to serve a six-month trial as a roughneck; Mobil didn't trust anyone in the office who hadn't first worked on the rigs—a wise policy that none of the oil companies upholds today. And what rigs those were; Bakersfield was one of the oldest working oil fields in the country, dating back to the 1890s. We worked on ancient wooden derricks and ate lunch in shacks where, in place of refrigerators, six-foot slabs of ice kept our food chilled. At the same time we were among the first to use steam injection to force out the last bit of residual oil; afterward the sands that held it would be scrubbed white. We were both behind the times and ahead of them.

I was one of the guys you'd see covered from head to foot with oil; I enjoyed the work immensely. And though I didn't realize it when I took the job with Mobil, I was exempt from the draft: Oil production was considered vital to national security.

Still I feared my petro-deferment would not last. The draft scheme changed all the time; we watched young couples cruise through Bakersfield on the way to Las Vegas to get hitched before the government stopped granting marriage deferments. Each month I wrote to let my draft board know I was still interested in further education and that I was taking night classes at Bakersfield Junior College.

I knew I wanted to pursue graduate studies but not what I wanted to study, other than something besides mechanical engineering. I gravitated toward two possibilities, nuclear engineering and oceanography. Susie settled on one, library science, though she would never finish the degree. (Instead she worked to keep me in school and out of Vietnam, something that's dogged my conscience ever since.) I applied to four state universities that offered all three disciplines—Oklahoma, Michigan, Washington, and UC–Berkeley—and was accepted by all of them. Looking more closely, I saw that the University of Washington was strong in all three disciplines. So in December 1965 we loaded all our belongings into our red Nissan Patrol and drove north. Come January, when I visited the departments, still undecided, physical oceanography won out over nuclear engineering. It just looked like more fun.

This time it was not just oil but the founder of Standard Oil and granddaddy of all oil men, John D. Rockefeller, who opened the doors for me. In 1927, the National Academy of Sciences reviewed the state of ocean science in North America and found that, with just 124 working oceanographers, this country lagged far behind others. The Academy recommended a number of reforms, including the establishment or expansion of oceanographic institutions at the nation's coastal corners. The next year Rockefeller put up $3.5 million—a fortune then—to implement many of the suggestions. As a result, Massachusetts got the Woods Hole Oceanographic Institution. The Scripps Institution in La Jolla, California, got some new buildings. And in Seattle the University of Washington got an oceanography laboratory.

The university's grant stipulated that the new lab be headed by a chemistry professor named Thomas Gordon (Tommy) Thompson, the first member of its faculty to be inducted into the National Academy of Sciences. In one of his first faculty appointments, Thompson recruited a gifted former student named Clifford Barnes, who was then languishing as a combustion engineer in Ohio. (The Depression had by then set in, and scientists took whatever work they could get.)

Barnes was still there in January 1966 when I arrived at what was now the Oceanography Department. But I didn't work with him at first; instead I started out under Maurice Rattray, a celebrated marine theoretician. Things didn't click; theoretical oceanography seemed arid and empty, far removed from the reality of the sea. And I was not a gifted theoretician. I needed a more hands-on approach.

That wasn't the only reason I struggled at UW. The curriculum then was brutally demanding, consisting of a year in each of the four disciplines of oceanography: geology, biology, chemistry, and physical oceanography. To be admitted into the doctoral program, a student then had to stand an eight-hour exam in each discipline—four days of exams, given once a year. Each student was also expected to pass exams in two foreign languages (Russian and French for me, for reasons I no longer recall). Furthermore, in the first two years of grad school, we had to take core courses in geology and biology plus six courses in advanced mathematics in the School of Aeronautical Engineering.

We students who'd transferred from other disciplines also had to take the senior-level classes in each of the four sequences. Those who didn't get at least Bs in these sequences were bounced out. For me, that would have meant Vietnam; I'd already passed my draft physical. And because I'd arrived midyear, I'd already missed the first courses in all four sequences. I was behind the eight ball. I received a D in my first exam—in the biological sequence—thanks both to my lack of preparation and to the fact that the professor teaching it was fresh from Germany and spoke very broken English.

Much as I wanted to fit in at the Oceanography Department, I seemed to march to a different drummer. But my biggest problem was that I did not yet have a feel for how the ocean worked. And I have always needed a sense of the big picture before I can understand the details of a subject. In this, I feel like a throwback to the naturalists of the 1800s, with their more holistic approach to science.

Fortunately, I found a mentor who shared that approach, and who would both save my academic career and start me on the research trail I still travel today. Cliff Barnes, the bearer of the Rockefeller legacy, was very different from the usual somber run of professors. His raucous laugh would ring through the entire oceanography building, and his snoring was just as penetrating. He claimed he could saw twenty-five thousand board feet of lumber a night; those who camped with him on research trips had to sleep at least a hundred yards away. When Barnes hosted visiting professors in the lecture hall, we grad students would watch from the back as he dozed in the front row and the light from the lectern glinted off his bald head—hoping he would start snoring and derail the lecturer.

I took classes in physical oceanography from Cliff and admired his thinking. He evidently liked what he saw of mine through my exam answers.

One day I knocked on his door and asked directly if he had any work for me. (Mobil would later support me in my studies, but at the time Susie and I were struggling.) A few days later he signed me on to one of the contracts he received from the U.S. Navy to study the waters of Puget Sound.

From there, things clicked; I received holistic teaching at the highest level. While my colleagues sought to define ever more minute aspects of the ocean, I was able to seek a broader, more integrated understanding before proceeding to finer-grained research. I'd finally found the mentor and, through him, the direction I needed. Cliff Barnes was no chalkboard oceanographer; he was an ardent, longtime tracker of real-world waters. There are hurricane chasers, tornado chasers, ambulance and skip chasers; Cliff was a water chaser. Thanks to him, I would become one too—and find my calling on a fjord called Hood Canal, one of the three arms of Puget Sound.

The ocean's flotsam washes into catch basins, and my hero Columbus and I both found our sea legs on such basins—Columbus on the Mediterranean, I on Puget Sound. During the Middle Ages, Columbus's ancestors moved a good deal around the Mediterranean. Later, as an able-bodied seaman and officer, he sailed along the shores of the Atlantic from Iceland to Africa, puzzling at the flotsam that arrived from the other side of the sea. I in turn puzzled over the Pacific drifters that washed into Puget Sound—my first inklings of a floating world beyond.

The University of Washington encouraged oceanography grad students to select a dissertation topic during their first two years in the program. Cliff Barnes, now my adviser, was interested in many subjects: icebergs, ice islands, Puget Sound, the Columbia River, and the little-known aqueous structures known as "water slabs." These water bodies-within-bodies are distinguished from the surrounding waters by slightly different temperatures—from a tenth of a degree centigrade to one degree—and hence different salinities and densities. They travel through the water as clouds float through air.

Water slabs were first detected by European oceanographers, who lowered water bottles and discovered varying temperatures at different depths. Barnes's student Ron Kollmeyer had identified the first local water slabs in

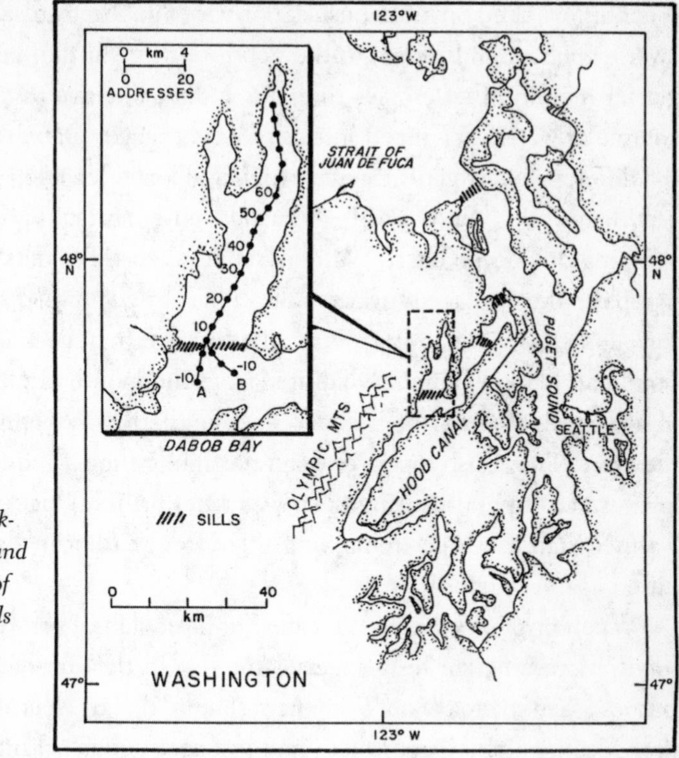

Dabob Bay, a backwater of Puget Sound and a microcosm of the ocean, with sills and transit points where snarks were sampled shown in inset.

Dabob Bay, a sub-fjord off Hood Canal. Kollmeyer used the metal vessels known as "Nansen bottles," set at regular intervals along a wire that's lowered to take water samples at precise depths. But the process is laborious and slow, and Kollmeyer could sample only once a month at nine stations spaced along Dabob Bay's ten-mile length—hardly enough to get a dynamic, comprehensive picture of the water's movements. It was like looking at amoebas under a microscope without adjusting the focus.

Nevertheless, Kollmeyer's explorations revealed the outline of a large water slab in Dabob Bay and suggested that slabs would likely form there in late summer and fall. During that season the bay's waters divide into three flow layers; the top and bottom layers tend to move one way and the middle one in the opposite direction. In summer the prevailing winds blow from the north, pushing water in the shallow uppermost layer to the south, out of the bay. Water in the second, middle layer flows into the bay to replace this outflow. This in turn pushes out water in the third and lowest flow layer, which extends down to the depth of the "sill," an ancient terminal glacial moraine that crosses the bay's mouth like a threshold. Often the volume of water exchanged in this process exceeds Dabob's "tidal prism," the difference between successive high and low tides.

Meanwhile, these northerly winds push surface water back from Washington's ocean coast. Warmer, saltier water then wells up from below to replace the displaced surface water. This upwelled water slowly makes its way along the sea bottom through the Strait of Juan de Fuca and into the Sound and Hood Canal. By fall it reaches the mouth of Dabob Bay. Most of it bypasses the bay, but every few days southerly winds appear and reverse the flows. They push top-layer water north into the bay, squeezing middle-layer water out and causing deep water to flow in. This inflow is slightly warmer than the water remaining there from earlier in the year. Vive la différence. A slab forms.

When my chance came to search for these slabs, I had the advantage of much more efficient testing gear. One day in the summer of 1967 Cliff Barnes asked me to try out some new, state-of-the-art electronic equipment the Oceanography Department had just acquired: a salinity-temperature-depth profiler (with the unfortunate acronym STD), which could read out temperature and salinity every ten centimeters or so as it was lowered into the water. STDs had not yet been tested in Puget Sound, and the department was eager to see how they would perform. Cliff loaned the STD and an old tugboat, the research vessel *HOH*, to me. (I never learned whether its name derived from the Hoh River on the nearby Olympic Peninsula or, more likely, from the formula for water, H_2O.) The *HOH* had a fantail—an aft working deck—close to the water, which was ideal for raising and lowering the STD. Thus equipped, I could sample at twenty stations along the bay, at every meter of depth, many times in a matter of days.

Even so, this first crucial test would have failed but for my engineering training. It was a technical trial in two ways. I planned to take along not only the salinity-temperature-depth profiler but a newly designed seven-track Kennedy tape recorder, to record data for later reading at the university's academic computer center. No one at UW had ever recorded oceanographic data this way. I tested the scheme first, dunking the STD at the dock and recording briefly. But the university's computer could not read the tape; the operators said it had the wrong electric parity. So back to the *HOH* I went and reset the recorder's parity switch, which was located behind a hinged panel. I dunked it again and returned to the computer center. Still it was the wrong parity. I turned the switch again, performed another test, and received yet another rejection. Round and round I went, resetting the parity each time. Finally, I watched carefully as I closed the panel and saw to my chagrin that an inconspicuous plastic bevel nudged the switch to the wrong position when the door shut. I cut it off with a hacksaw. One last test and the computer center could read my tape. If I had not persisted I would have recorded unreadable data all weekend, disappointed Cliff, and wound up who knows where today.

Dabob Bay snarks in late summer. The horizontal view is highly compressed to fit in the frame. The snarks, shown here as rounded black blobs, are actually thin slabs more on the scale of the arrows. A through G represent various flow features: A, surface water driven northward by winds from the south; B, a region of rapid change in water density known as a pycnocline, *which separates the surface layer from the out-flowing water below (C); D, a "tongue" of stagnant water at the bay's head; E, a deep inflowing layer segmented into snarks by pulses of southerly winds; F, another region of rapid change in water density separating the inflowing snark layer from water (G) trapped below sill depth.*

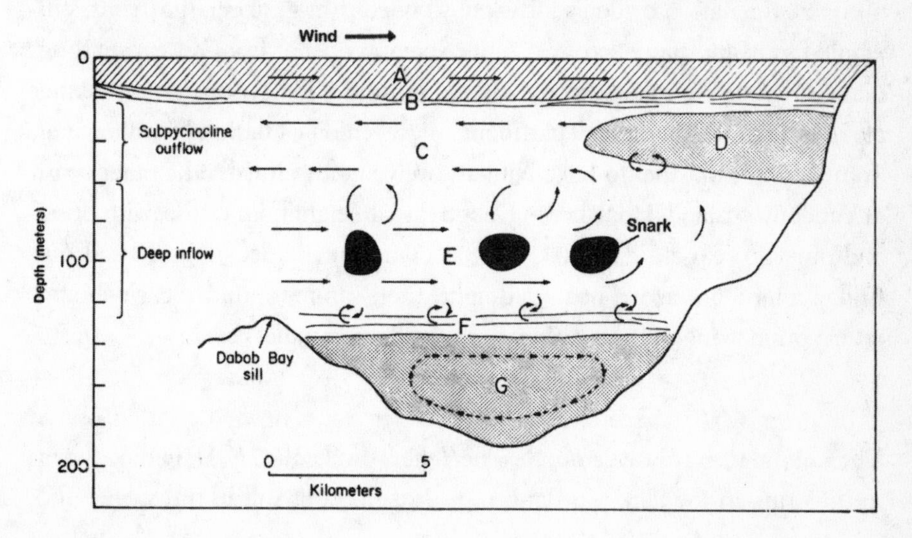

As it was, the first Dabob results proved spectacular. As soon as I got back from this, my first oceanographic expedition, I copied the readings onto a grid and drew contours of the varying temperatures, just like the elevation contours on a topographic map. These contours, appearing as concentric rings, revealed a rich tapestry of slabs. I worked round the clock for a weekend and brought back seven transects of Dabob Bay. They showed one slab moving about one mile north and south with the tide. I later found that once the slabs became identifiable as distinct entities, they would last for a week or more. (Later still I would learn that in the ocean they could remain distinct for years, long enough to cross the North Atlantic at a depth of several thousand feet.) I traced the contours onto vellum, colored them with brilliant inks, and presented the results to Cliff. Impressed, he had the vellums made into slides for his lectures—slides I still use in PowerPoint presentations today.

Thus began my oceanographic field work, in the infant science of water slabs. The science was sound, but its whimsical appeal was inescapable. My daughter, Wendy, liked to call them "water packages," but I named them "snarks" in honor of their elusiveness, after the "inconceivable creature" of Lewis Carroll's poem "The Hunting of the Snark."

I knew I had the topic for my dissertation: the hunting of snarks in Dabob Bay. But getting time on the water could be difficult; the department's resources were stretched thin by a surge of entering students. To handle the overload, the navy provided a forty-foot wooden river gunboat, the *Tenas*, still painted gray and propelled by a single-screw propeller from an ear-splitting diesel engine. We grad students ranged far and wide on the *Tenas*, taking along our families and the department's diesel credit card. Bob Hamilton and Tom Hopkins took her to Lake Nitinat, twelve hours away on the outer coast of Vancouver Island. I took her to Dabob Bay, an eight-hour cruise, with Susie and our two-year-old daughter, Lisa, sleeping on the deck. We'd head out Friday afternoon, arrive near midnight, then sample round the clock and return Sunday night. I was twenty-four and had boundless energy.

The sailing wasn't always smooth. The *Tenas*, unlike the *HOH*, had no radar, so motoring in the dark was dangerous. Logging was still in full swing, and countless deadheads floated around the Sound. Returning one night, I foolishly throttled up to full speed and hit one. Luckily, it only bent the propeller's drive shaft, and we limped home. On another trip, I mistook the lights on a ring of tribal fishing nets and wound up trapped in the middle of them. Great deck lights flashed and blinded me, and I feared I'd be shot. It took hours of painstaking effort to extricate the *Tenas* without tangling her prop in the nets.

Sailing on Hood Canal could entail even stranger encounters. Bangor, right across the canal from Dabob Bay, is home port for the Trident submarine fleet, the backbone of the United States' nuclear arsenal. I often steamed past the subs; they used Dabob as a torpedo testing range. At its mouth a traffic-style light indicated whether they were firing: Red meant "stay out." Green meant we could enter and chase slabs. Once during heavy fog the radar on the *HOH* went out just as we passed Bangor and we had no choice but to put in. The navy ordered us via radio phone not to dock, but the fog was so thick we had no choice. A radar expert appeared in a flash and within minutes fixed our equipment so we could shove off. I had never seen a technician so fearsomely competent. It's enough to give you confidence in our nation's defenses.

Later, the navy would come to take a keen interest in water slabs, because they deflect sound waves and thus provide spaces where submarines can hide from sonar. They are, or should be, of concern to everyone who deals with the diffusion of sewage, spilled oil, and other pollutants in water, because they slow that diffusion. But such practical impacts did not interest me. I found snarks fascinating, even beautiful, in their own right.

In the summer of 1968, I mounted an intensive series of field trips, taking the *Tenas* to Dabob every two weeks or so. I found three snarks in a row—produced by three intervals of south winds—and discovered that surprisingly weak winds could cause them to form a full three hundred feet below the surface. But fortnightly snapshots could only reveal snarks after they'd coalesced. I wanted to see the process at work—to see a snark actually form. It wasn't easy; access to the ship and STD was tight, handling the *Tenas* was demanding, and it was difficult to keep up with course work as well. But by September I could see the snarks building to a crescendo. I was a man on a mission, and the whole faculty knew it.

Fortunately, a well-placed angel was looking after me. Though I didn't know it at the time, I was Cliff Barnes's golden boy, and I suspect he put in a good word with a colleague at the federal Bureau of Commercial Fisheries. For some reason I never understood, the bureau offered me a 212-foot-long research vessel for a week, and I took advantage of the offer two weeks later. The faculty gaped when it docked outside the Oceanography Department at the service of a lowly third-year grad student.

We arrived at Dabob after a half day's steaming, just as a huge snark entered the bay. As luck would have it, the light was green; no submarines were testing torpedoes that week, so we could charge quickly up and down the bay. On each transect, we saw the giant snark pushing farther in. At first it moved in normal fashion, but then a mass of denser water followed behind it and erupted into a flurry of snarks. These sank into the stagnant basin, displacing water that was still sitting there from the previous year. I'd been doubly lucky. Not only had I observed new snarks forming, I'd seen them displace older water below the sill.

I discovered that because the winds blew episodically, water flowed into the bay in bursts. This injected a series of snarks, like links squeezed into a sausage casing. During quiescent intervals the snarks drifted to the head of the bay, where they eventually welled up and disappeared. In this relatively calm environment, a snark can survive for weeks, and it can take two years for diffusion to erase the signature of water trapped below the sill depth.

Slight variations in density and current produce very different snarks. I have never seen snark shapes repeat themselves; like snowflakes, each one is unique.

After one summer chasing snarks, I had what I needed for a dissertation. And I'd gained a new sense of how water works, not just in this little fjord but around the world. Dabob Bay, two hundred meters deep, is a natural laboratory for the study of the sea; conditions there verge on those of the oceans' middle depths. It turns out that Mediterranean eddies or "meddies"—slabs that form when dense, warm, deep water exits the Strait of Gibraltar—behave much as Dabob snarks do. Like snarks, meddies can drift long distances, as far as Bermuda, without breaking up.

We commonly think of water as homogeneous: one molecule of H_2O after another, each the same as the last. And indeed, most oceanographers see the sea that way. This is called a Eulerian view, after the eighteenth-century mathematician Leonhard Paul Euler, who advocated observing the universe from fixed points of reference. But water bodies are also granular, composed of distinct, unique entities—slabs that chart their own courses. I've always tended to see water this way—as a collection of dynamic elements. This is called a Lagrangian view, after Euler's colleague Joseph Louis Lagrange.

Imagine two police officers monitoring traffic on a highway. The Eulerian cop stands beside the road, tallying each passing car's speed and position at a single instant by radar. The Lagrangian cop cruises within the traffic, tracking a single car's progress over a period of time. Both viewpoints contribute to our understanding, and both have their limitations. We Lagrangian water chasers offer a necessary corrective to the Eulerian conventional wisdom.

Water bodies, like human bodies, contain discrete parts. Slabs, snarks, and meddies travel through them just as food, air, and blood course through living bodies—or as colored blobs rose and fell in the heated Lava lamps that were so popular in the sixties, whose liquid appeared homogeneous at room temperature. To visualize the ocean, imagine a vastly larger, flatter lamp, with the blobs stretched sideways and time slowed to a snail's pace. If each slab were a different color, the ocean would look like a Pointillist painting. Imagine these slabs—or dots—pushed about by the wind, colliding and separating; as Cliff liked to say, it's easier to move slabs around than to mix them. The contours they form resemble knots in wood. I can hardly look at a cross section of an old tree or the knotty-pine paneling in a beachside motel without recalling the temperature contours around snarks.

"It's a little ocean over there," Cliff would say, as he encouraged me to unravel Dabob Bay's secrets, and he was right. With its slow currents and multilayered flows, Dabob is a marine microcosm, the whole ocean in a small, accessible frame.

2. Oil and Icebergs

You must let yourself go along in life
like a cork in the current of a stream.
—Pierre-Auguste Renoir

Looking back on my student days, I'm amazed how many connections formed in the University of Washington's Oceanography Department led to vitally important collaborations in years to come. One in particular would only bear fruit twenty-five years later. Jim Ingraham was four years older than I was and already working in the oceanography graduate program when I started there in 1966. We both held oceanographic jobs while we attended classes; I did part-time research for Mobil and the U.S. Navy and served as Cliff's teaching assistant, and Jim worked full-time at the federal Bureau of Commercial Fisheries (now the National Marine Fisheries Service), right across the interlake canal known as the Montlake Cut from UW. We were both fortunate enough to work under pioneers of marine science: UW's Cliff Barnes in my case and, in Jim's, the formidable (and happily named) director of the Fisheries Bureau's oceanographic program, Felix Favorite. We both stood on the shoulders of giants.

Back in late 1962, as he was preparing to complete his master's in oceanography at UW, Jim trudged across the Montlake Bridge to the Fisheries Bureau's lab to interview for a temporary job. That job became permanent when another federal oceanographer crossed the bridge the other way to take a job under Cliff Barnes. It was a close-knit profession.

Two weeks later, Jim was on his way to the Aleutian Islands to join Felix on a joint fisheries-oceanography "cruise" (oceanography jargon for an expedition) on the same federal ship that would later enable me to complete my snark hunt in Dabob Bay. At that time, the eastern North Pacific was *aqua incognita*, a vast expanse largely absent from the annals of oceanographic data. It's also the sea that various salmon species—the Northwest's staple fish—orbit as they grow to maturity, before returning to their home rivers to spawn. (These migrations last about as long—two to four years—as one revolution of the North Pacific Subarctic Gyre, which reaches from southeast Alaska to Siberia. That's probably no coincidence.) Favorite sought to learn what marine factors influenced the salmon's choice of offshore winter grounds. More cruises followed that year and the next and the year after that. Future analyses of the data they collected would show how the salmon followed oceanic pathways defined by sharp changes in temperature and salinity—the same factors that delineate snarks. And sure enough, these cruises traced snarks orbiting the gyre.

In 1965, back at the Seattle lab, Jim started on a breakthrough project in computer simulation that would much later prove to be a scientific lifesaver. The computers of the day were primitive, but the Fisheries Bureau scientists were innovators in their use. They were among the first to deploy a programmed data processor from the Digital Equipment Company—then the state of the art—at sea. This enabled them to process all their data on ship rather than laboring for months afterward at the office. But they needed a way to evaluate the effects of ocean currents on the migration of salmon. At the instigation of his brilliant supervisor, Taivo Laevastu, Jim developed a computer program that crunched the speed and direction of the currents and salmon.

This was the beginning of the Ocean Surface Current Simulator, OSCURS for short. OSCURS would add a whole new class of oceanic drifters—floating indicators of the sea's movements—to the water chaser's toolbox. Already, with the snarks, I had been tracking natural drifters. Soon

I would begin using determinate, or deliberate, drifters—message bottles, barrels, drogues, and all the other objects oceanographers have released and beachcombers have retrieved for two centuries. In the 1990s, I would discover how much accidental drifters can tell us about the waters' ways.

Jim meanwhile would create mock computer drifters that could interpolate complex paths across thousands of miles and many years, even decades, at sea for objects observed at intermittent, isolated positions. With OSCURS, he could predict long drifts from just a few known coordinates.

I did not get to know Jim well at UW; we were both too busy at our studies, research, and jobs. But I knew him well enough to call him up after the shoes dropped.

I left the University of Washington in 1969 to work full-time in New York City as Mobil/Standard Oil's first oceanographer. It could have been a lonely, frustrating post, but Clare J. Colman, the chief of Mobil's offshore operations, was a kind and tolerant manager who knew enough to give wide latitude to a solitary oceanographer in a crew of engineers—though sometimes I pushed even the bounds of his tolerance.

Other new UW oceanography grads also fanned out into the oil industry. One, Doug Evans, who had been Cliff Barnes's teaching assistant for my classes, finished his master's degree in 1968 and went to Shell Oil in Houston. Another, Bob Hamilton, completed his master's in 1967, joined a consulting firm in Anchorage, and then, in 1968, came to Houston to work for the Baylor Corporation, which made wave staffs for measuring wave heights. Rockefeller's gift forty years earlier had borne some fruit. Our success in the private sector also testified to Cliff Barnes's practical, hands-on approach to teaching. Nevertheless, Cliff seemed disappointed. "I thought you'd stick around for a few more years," he told me just before I moved to New York. But I had a growing family, and I needed a job.

We arrived in New York in November 1969, just after our second daughter, Wendy, was born. While we were still settling into our apartment in Scarsdale, I was asked to quickly help design a drilling platform for remote Sable Island, about ninety-seven nautical miles east of Nova Scotia. I worked the weekend and came up with a deck high above the sand. Up went my report to Mobil's vice president for engineering, who asked in bewilderment why it needed to be so high. I explained that it was to protect against storm surges: When I added a hundred-year surge to a hundred-year tide, I reached a maximum wave height well beyond what were then the benchmarks in marine construction. He sent me to Sable Island for a firsthand look.

In 1970, I flew in a Grumman Widgeon from Halifax to Sable Island, a boomerang-shaped, twenty-three-mile-long sandbar near the junction of the North Atlantic's two great circulating gyres. This location makes it a marine graveyard; shipwrecks and other debris from both gyres wash up there, and we landed amid a gaudy litter, the sand nearly burying the wheels. It was the first such wash-up beach I'd ever seen, or at least noticed, though such beaches would later prove to be vital sources of drift data. I brought home one artifact, a fishing net—nearly the first beachcombing I had ever done.

Later that year, I represented Mobil in an intercompany effort called the Ocean Data Gathering Project—in effect, a conspiracy of the UW oceanography grads who'd gone to work in the oil industry. Our goal was to challenge the prevailing industry wisdom on the threat hurricane waves posed to oil rigs in the Gulf of Mexico. That dogma pegged the hundred-year wave—a wave with a one-in-a-hundred chance of occurring each year, the standard the oil companies built for—at fifty-five feet. But we believed Gulf waves could rise much higher, with terrible consequences. We needed data to support our contention, so we dreamed up the Ocean Data Gathering Project.

The project contracted Bob Hamilton to equip six platforms in the Gulf with wave staffs—pairs of wires stretched vertically to a height of one hundred feet—to electronically record exactly how high the waves rose. Bob was a genius at rigging offshore platforms with instruments, and he and his staffs got the goods. Hurricane Camille passed within fourteen miles of one platform off the Mississippi Delta, and his instruments recorded a seventy-two-foot wave. We had made our point, and the industry design standard was raised to protect against waves of up to seventy-five feet. That standard remains in effect today—although recent measurements, such as a ninety-five-foot wave recorded during Hurricane Ivan, suggest it should be even higher.

For the next five years, I traveled incessantly for Mobil, studying Sable Island and the icebergs that threatened the Hibernia oil fields 170 nautical miles southeast of St. John's, Newfoundland, on the outer Grand Banks. Every couple of months I flew from New York to Mobil Oil Canada's offices in Calgary to present these studies. On each trip I'd stop for a few days in Seattle, visit Cliff Barnes, and work on sorting out my snark data. This peripatetic schedule actually helped me finish my dissertation and get to know him much better than I would have if I'd stayed on campus.

When I visited, I slept in Cliff's basement, amid his fly-tying gear. (He was a renowned fly fisherman with a collection of books on the art dating back to the 1600s.) We worked weekends on the snarks and I helped him with his ongoing Puget Sound projects; the fact that he sought my advice seemed the ultimate compliment. In the evening we would sit in front of the fire and he would bring out his favorite refreshments, Olympia Beer and Planter's dry-roasted peanuts. Cliff was a natural storyteller with many tales to tell. He talked about fishing and how he grew up on a farm in Goldendale, Washington, rode his horse to a one-room schoolhouse, and hated farmwork. And he recounted the strange coda to his wartime service—how in 1946 he'd been ordered to rush straight from Argentia, Newfoundland, to Bikini Atoll to witness the first H-bomb test, still wearing his Arctic parka.

During the war, Cliff tracked icebergs in the waters that claimed the *Titanic*, in a life-and-death struggle to save convoys bound for Britain from submarine attacks. The convoys steered through the jaws of a vise. To the south, U-boats waited to pick them off. If they strayed too far north they'd be pulverized in Iceberg Alley by some of the ten thousand or so icebergs calved off Greenland each year. Somewhere between, determined by ocean currents, lay the sweet spot. It was Lieutenant Commander Barnes's job to help find it.

At Argentia, Newfoundland, Lieutenant Commander Clifford A. Barnes calculates iceberg threats to wartime convoys, May 1945.

I listened eagerly to his tales; icebergs were a subject of more than passing interest for me. One of my first assignments at Mobil was to represent the company in a Memorial University study on whether icebergs might be rerouted to keep them from colliding with offshore oil platforms, which would determine whether it would be feasible to extract oil from the Grand Banks. The university pitched a scheme to have large oceangoing tugs—usually two or three per iceberg—lasso a berg and tow it ever so slowly and slightly off its collision course.

Since then it's become routine to tow icebergs weighing 250,000 tons, half the size of the one that sank the *Titanic*; a 7.5-million-ton berg has been successfully towed. But it was a novel notion then, and Cliff shared sobering tales about the difficulties of foiling icebergs. Many a wartime mariner owes his life to this humble farm boy who spotted the bergs first. At least once, however, the ice outwitted him.

One evening, while a four-foot-long caiman that his daughter, Janet, was petsitting napped by the fireplace in incongruous tropic splendor, Cliff told me about that terrible time. He was performing reconnaissance in a PBY (patrol bomber) antisubmarine plane. A hundred miles from the *Titanic*'s grave, he spotted a berg forty-five hundred feet long and thirty-six hundred feet wide, rising sixty feet above the surface and grounded in six hundred feet of water. Then the infamous Newfoundland fog closed in. When it lifted, the ice mesa had broken into an armada of shards, floating right in the convoy's path. In the ensuing confusion, about sixty ships collided with each other trying to avoid the bergs. As far as I know, none sank.

What Cliff found, then lost, that foggy wartime night was no ordinary Greenland iceberg. It was one of the wide, flattened bergs known as tabular icebergs. Rather than calving from Greenland glaciers as conventional icebergs do, tabular icebergs typically break off from the shelves of sea ice that extend from Ellesmere Island, northwest of Greenland, onto the Arctic Ocean. They have about the same length-to-thickness ratio as water slabs, and, like oceanic slabs, they can be prodigious in size.

The Arctic peoples have long known about the ice islands' movements and the oceanic conveyor belt that propels them. In the 1920s the Danish explorer Knud Rasmussen recorded this verse by an Igloolik Inuit woman shaman named Uvanuk:

The great sea
has sent me adrift;
it moves me
like a reed in a great river.

Earth and the great winds
move me, have carried me away,
and filled my inner parts with joy.

Irish monks may have been the first Europeans ever to sight an ice island. By 795 CE, sailing in leather boats, the monks had crossed the northern sea and reached Iceland. When the first Vikings arrived around 874 CE, they were surprised to find a monastery there. Three centuries earlier, according to the medieval *Voyage of St. Brendan,* the Irish monk Brendan of Ardfert sailed his leather boat even farther and saw a "pillar in the sea" so large it took three days to come up on it,"so high that Brendan could not see the top of it," whose foundation extended "down into the clear water."

Thirteen centuries later, steel and wooden ships had replaced leather boats, but Arctic exploration was scarcely less perilous. In October 1872, in the narrow passage between Greenland and Ellesmere Island, a crack in the ice separated nineteen people, including five women and children, from the U.S. surveying ship *Polaris*. Trapped on the floe, they drifted south through the Arctic winter, some seventeen hundred miles in 198 days. Miraculously, all survived. A sealing steamer finally rescued them.

Much larger ice islands than the ones Cliff Barnes and St. Brendan spotted have haunted these channels. In October 1883, the great anthropologist Franz Boas observed a three-by-eight-mile island in Cumberland Sound, between Baffin Island and the Canadian mainland. In 1918 Storker T. Storkerson, a Norwegian explorer temporarily heading the Canadian Arctic Expedition, set out on a bold gambit: to ride an "ice cake" from Cross Island, north of Alaska, to Siberia, following the westward current that was then believed to circle the Arctic Ocean.

Even before that, Jules Verne had imagined islands—both natural and man-made—floating across the sea. In his 1873 novel *The Fur Country,* the earth beneath a trading post gets unmoored by an earthquake, then locked in sea ice through the winter. Its inhabitants must await the spring thaw and pray that the "Bering current" will bear them back southward. But Storkerson and his four companions appear to have been the first brave souls ever to deliberately set out to ride a floating island. They drifted 440 miles in 184 days on a floe he estimated to be "seven miles wide and at least fifteen miles long," with contoured hills "in which numerous small lakes and ponds were visible." Its "ridges and levels" were, he recounted, "exactly like certain stretches of prairie."

Storkerson and his comrades took six weeks' worth of provisions and consumed just two weeks' worth, subsisting quite happily on seals and polar bears. "So far as we could judge we could have lived on the ice island eight years as easily as eight months," he reported cheerily. They proved that Arctic expeditions could live off the land, or ice, and that no westward current circled the Arctic. Instead of reaching Siberia, they traveled the gyre that circles the Beaufort Sea, which we have chosen to call the Storkerson Gyre in honor of his feat.

Some quasi colonizers have since stayed even longer on ice islands. In the 1930s, the Russians began manning Ice Island SP–1, which drifted from the vicinity of the North Pole to northern Greenland. In August 1946, three hundred miles north of Point Barrow, American air-reconnaissance patrols discovered an arrow-shaped, 140-square-mile chunk of floating ice ten times thicker than pack ice. Its surface was corrugated with icy waves, like Storkerson's "rolling prairie," in sharp contrast to the fractured-mirror look of the surrounding ice. The crests rose twenty-five feet above the troughs.

Despite this island's size and proximity to Alaska, the U.S. military managed to keep it and two other ice islands secret through the cold war. Once the military had determined they were stable, it began basing operations on them. It dubbed the ice islands "targets" because of their distinctive appearance on radar: Target 1, or simply T1, and so on. T1 drifted west to Barrow, then north along the Storkerson Gyre toward the North Pole, and finally ran aground on Greenland. T2 started on the same route, then exited the Arctic via another current, the Transarctic Stream, and disappeared. But T3 continued drifting around the gyre for nearly three decades. A generation of oceanographers dwelt and conducted research there. Other researchers took a botanical inventory and found tree branches, caribou antlers, large boulders, and a tussock of living moss. Tree rings revealed that T3 had broken off the Ellesmere Ice Shelf sometime between 1936 and 1947. The moss suggested that ice islands could be vehicles for dispersing plant species around the Arctic.

Finally, after looping twice around the Storkerson Gyre, Target 3 also escaped on the Transarctic Stream. Pieces of it eventually reached the southern tip of Greenland. By 1984, what was left of T3 had traveled eight thousand miles—the longest continuously documented drift of any floating object.

Most icebergs melt when they reach what we have dubbed the Titanic Terminal, the point at the southern end of the Grand Banks where Iceberg Alley smacks into the warm Gulf Stream. But some drift much farther, though melting drastically shrinks them. Some have crossed the Atlantic to Britain. In June 1926, a fifteen-by-thirty-foot chunk passed Bermuda. It may have originated in the Arctic Ocean, more than four thousand miles away.

Icebergs weren't the only threat to offshore oil platforms. In the early 1970s, Mobil and other companies were hungry to exploit the rich petroleum deposits discovered under the deep waters of the North Sea, between Britain and Norway. But what sorts of waves would platforms have to withstand there? Fishermen had reported enormous crests, but official oceanographers for the most part ignored these reports as mere "anecdote." We had none of the oceanographic data needed to support engineering designs. So I coordinated some early studies, which suggested monstrous waves did indeed strike the North Sea—ninety-five feet from trough to crest, an assault only a very large structure could withstand.

Mobil led a consortium of oil companies in the effort to design such a structure. The project's lead engineer was Frank Manning, a member of my offshore group and a brilliant problem solver with more than twenty-five patents to his name. I oversaw the project's environmental aspects. The stakes were high. On one trip to Norway, I visited an enormous oil rig that had been damaged in a storm and was now in the shipyard for repairs. Such an incident could easily tip into a human, financial, and environmental disaster.

We put out requests for proposals and received predictable responses—designs for the sorts of steel structures that were standard at the time. But a firm named Norwegian Contractors proposed a radically novel design that came to be known as Condeep (for "concrete deep water"). A dozen or so oil-storing silos with concrete walls up to several feet thick would rest on the seafloor, four hundred feet down. Three of them would rise a hundred feet above the surface, supporting the platform. We bucked tradition and selected the Norwegian design for what became the first of seventeen concrete structures erected in the North Sea for about $300 million (in 1975 dollars) apiece. Our caution seems to have paid off: Since then, at least one ninety-five-foot wave has hit a North Sea platform without serious damage.

To support this decision, I had model tests run in the huge wave tank at Trondheim University. An elaborate system of powerful, electronically controlled paddles subjected a model of the Condeep structure to the same kinds of forces that would attack the actual platform, and instruments measured their effects. This was cutting-edge engineering and I worked closely with Norwegian scientists and the risk management firm Det Norske Veritas, who took a close interest in such issues. The results were encouraging, but I cross-checked them by having a computer simulation run at the University of California at Berkeley.

Between these tests and London meetings, I was forever flying to, from, or between Britain and Norway. It seemed a waste to jet halfway around the planet just for a business meeting, so I'd add a few days to each trip and visit museums, beaches, piers, and friends. In Oslo, I saw Viking ships and the nineteenth-century polar exploration ship *Fram*, which drifted halfway across the Arctic Ocean. In Amsterdam I examined the dikes. Twice I stopped off in Bergen to visit August Faye, the father of Doug Evans's wife, Randi. Though I did not know it at the time, back in the 1840s August's great-grandfather, Christopher Faye, had invented the glass fishing float, which launched the industrial fishing era by making it possible to deploy much larger, more efficient fishing nets. In the decades that followed, Norwegian glassblowers would produce about 16 million of these glittery floats, many of which would tear loose and circle the North Atlantic. Japanese fishermen would deploy 120 million floats in the North Pacific. Some still wash ashore today, providing valuable clues to the movements of the gyres.

In 1972, the Mobil Research and Development Corporation decided to move its offshore engineering section away from New York. It considered two possible locations: Princeton, New Jersey, and Duncanville, Texas, just outside Dallas. Susie got down on her knees and prayed for Princeton, but Mobil settled on Duncanville. So we headed for Texas.

Life in Dallas had one consolation: I got a taste of the academic life I had abjured when I went to work in industry—teaching an evening oceanography class at Southern Methodist University, just a block from our apartment. Soon after I started, my boss, the ordinarily easygoing Clare Colman, told me that such moonlighting was a breach of Mobil policy. I said I would be happy to return the $300 I'd received and tell SMU that Mobil objected to my teaching. Mobil relented and I continued trying to bring the ocean to landlocked Dallas for two semesters.

The class attracted some interest, and soon a local TV station asked me to discuss oil spills on a Sunday morning public-affairs program. I assumed Mobil would not object to such an unpaid appearance. I was wrong again; managers at the highest levels went ballistic. I told Clare that I would gladly tell the station it was against Mobil policy for me to speak. Mobil, evidently fearing even more embarrassment, relented, but a member of its board of directors came to Dallas to tape my interview. All went well, and I managed not to say anything that would get me fired.

This was my first contact with the politics of oil, but I didn't have any strong feelings on the subject. Oil had not become quite so freighted a subject as it is today.

But though I bore no grudge from the dustup over my TV appearance, I realized my days at Mobil were nearing an end. I still could not get enough time at home; I was out of the country at least a third of the time. These trips could be instructive and exhilarating—I managed to visit most of the seven oceanic gyres in the Northern Hemisphere. But my absence was hard on Susie and our daughters, and Dallas was wearing thin for all of us. I realized I had to get out.

Through the years, I'd kept in close contact with my old classmates Doug Evans and Bob Hamilton, who had recently formed their own marine consulting firm in Houston and Washington, D.C. When Mobil sent me from New York to Houston for progress meetings on the Ocean Data Gathering Project, I stayed with Bob and his wife, Glenda. Now, with Dallas and the future weighing on my mind, I sounded him out.

Curt Ebbesmeyer at his Mobil office, Duncanville, Texas, ca. 1972.

I suggested we drive from Houston to Galveston—as it happens, a prime wash-up beach that would prove a treasure trove in my future beachcombing days. As we talked, we waded along the surf, towing a case of beer wrapped in burlap and suspended in an inner tube behind us. After we were well sloshed, I bared my soul. I told Bob that I would have to quit Mobil and leave Dallas. The timing was good; I'd just completed my dissertation and vested my retirement plan at Mobil. But prospects were grim for finding a congenial job in marine science; universities looked down on scientists who'd worked in industry, particularly the oil industry. And I knew I was not suited to the vicious politics and endless, often pointless committee meetings of academe; how Cliff stood it all those years I'll never know. The only avenue that seemed open was consulting, which the university types then scorned despite the fact that good research was done there. (Now they openly compete with consulting companies for the work.)

Meanwhile, Seattle beckoned. It would be a better place than Dallas to raise Lisa and Wendy. Cliff was there, and Akira Okubo, a friend and colleague from back East, also planned to move there in 1974. I told Bob I would return to Seattle, even if I didn't have a job, and strike out as a consultant.

Afterward, Bob conferred with Doug. "I'd rather have him working with us than against us," said Doug. They offered to stake me for a year if I would open the western regional office of their fledgling consulting practice. Once again I felt the shade of John D. Rockefeller pass by. Early in the century, Bob's grandfather and namesake, Robert Clark Hamilton, had founded the Clark Oil Company. Rockefeller decided to buy him out, as he did so many other competitors. He offered Hamilton $50,000 in cash or Standard Oil stock. Old Man Clark took the cash. If he'd taken the stock and kept it, his family's fortune would been made. Grandson Bob might never have had to work. And my professional fortunes might have turned out very differently.

In December 1973, seven years after I started grad school there, I returned to Seattle to take (and pass) my final PhD exam. Cliff took me to lunch at Ivar's Salmon House, a classic Seattle restaurant modeled on a native cedar longhouse, and celebrated with two double martinis. I was his last student; he had waited for me to finish my snark studies before retiring. I did not know it at the time, but he was coming down with Alzheimer's disease, the symptoms of which would become obvious a few years later. I would eventually inherit his research papers, and within them some startling revelations about postwar mines in the Pacific and Viking mariners, the first oceanic drift researchers, in the Atlantic.

The next March—back in Dallas at the peak of the Arab Oil Embargo—Susie and I packed our Volkswagen Squareback so full the girls had to lie stretched out on the stacked baggage and drove to Seattle. When we arrived, we found an apartment across the street from the single room we'd rented in 1966. But we'd set aside most of my Mobil retirement money for a down payment and, once we'd worked up the courage, set out house shopping. After just a day we found the 1917 bungalow we still live in today, a mile north of the university. We applied for a loan from a local bank, and the banker called my office (our apartment) to verify my employment. "Evans-Hamilton," Susie answered. "Could I speak to the manager?" the banker asked. She handed the phone to me and he asked whether a Curtis Ebbesmeyer worked there. "Yes, indeed," I replied. He asked what his salary was and I told him. He hung up without asking my name, and we had our loan.

As in our student days, we cut our expenses to the bone. I couldn't pay a secretary; I could barely afford an old IBM Selectric on which to type proposals for consulting contracts. So Susie worked for free. When she fell ill with the flu, I brought the typewriter to her in bed and we continued churning out proposals, desperately seeking work. One day we lost three bids in one morning and threw up our hands. We packed off to the nearby St. Michelle winery, whose tasting tours dispensed all the wine you could, ahem, taste. We tasted our way to blissful oblivion, went straight home to bed, and resumed the struggle in the morning.

Fortunately, I was a salesman's son, and I'd picked up Dad's knack for knocking on doors. One door finally opened, at the Municipality of Metropolitan Seattle, better known as Metro, the regional sewage agency. Metro wanted to expand its huge treatment plant at West Point, a beach jutting into Puget Sound from Magnolia Bluff northwest of downtown Seattle. To do so it had to show that the additional effluent would disperse into the Sound's currents. And to determine that, we had to parse the movements of water—i.e., water slabs—in the Sound.

The work was right up my line; once again I would be chasing snarks, though of a very different kind. The slabs that formed in the Main Basin of Puget Sound, with its powerful currents and surging tides, moved more than fifteen times as fast as Dabob Bay's relatively sedate slabs. I came to call them "hypersnarks."

I'd happened to visit with Alan Thorndike, a University of Washington physicist who was studying the behavior of sea ice on the Arctic Ocean. Thorndike explained how he and his colleagues set out markers on pack ice and calculated the rates at which they spread as the ice fractured and drifted. I thought, Why not use the same approach to decipher turbulence in water? I released drogues—underwater sails attached to floats—and calculated their changing positions every hour or so using the same tool, a sextant, that ancient mariners used to note the position of the stars.

Meanwhile, William P. Bendiner of UW's Applied Physics Laboratory injected effluent from the plant with fluorescent dye, which subsequently

Releasing drogues on a calm day off West Point, 1974.

became trapped within the hypersnarks. The movements of the drogues and dye dovetailed perfectly with that of dye injected into the hydraulic model of Puget Sound at the university. From all the data I determined that the snarks were formed by intense tidal mixing in the Tacoma Narrows—a turbulent choke point about midway in the Sound. Each tide would deliver a batch of water of slightly different density, and all these batches would then seek their appropriate levels in the Main Basin—the densest on the bottom, followed by the next densest, and so on. This produced the structure known as a "devil's staircase" in mathematics, a series of erratically sized steps such as a drunken carpenter might construct. Sewage became trapped in these layered hypersnarks and failed to disperse as rapidly as the standard mathematical models for continuous, uniform water bodies predicted.

Returning to Seattle proved fortunate in another way. It reconnected me with someone I'd gotten to know back East, a mathematician and oceanographer who would become my second essential mentor and collaborator after Cliff Barnes.

I'd first become aware of Akira Okubo in the course of my work on water slabs. I'd come across papers he had written using just the kind of mathematics—advection-diffusion from a point-source release of material—that I was trying to apply to the Dabob snarks. Okubo had trained as a chemical engineer at prestigious Tokyo University and, like me, worked as an engineer for a short time. But he soon tired of engineering and sought work that would allow more scope for his imagination and talents, which were prodigious. Give him the most complex, seemingly intractable problem and he would come back a few days later with a precise, elegant solution elaborated across page after page of neat mathematical formulae. So he turned to oceanography. The ocean is, after all, just a much larger, more complex vat in which to mix things, and he could apply the same mathematics he'd used to unravel chemistry problems.

Okubo wrote to Don Pritchard, head of Johns Hopkins University's Chesapeake Bay Institute and one of the leading oceanographers of the 1960s. Pritchard got him admitted and advised him on his thesis, which examined the role of currents' turbulence in a substance's diffusion through water. But even then, as I would learn much later from Pritchard, Okubo was looking toward the next horizon. He dreamed of using the mathematics of dispersion in water to figure out the behavior of icebergs, insects, and other seemingly random swarming phenomena.

In 1970, I tracked down Akira Okubo at Johns Hopkins and gave him a call. I described my interests and we had a pleasant, productive chat. To my surprise, the next week he came to visit me at my office in the Mobil Oil Building, kitty-corner from Grand Central Station. We had another interesting chat and I showed him the calculations I'd been working on. He seemed pleased that I was following up on his previous research, and I kept him apprised of my progress on my doctoral dissertation.

My friendship with Akira proved vital to completing my PhD. Maurice Rattray, now a member of my doctoral committee, said he was unfamiliar with the math I was pursuing, but if Akira would sign off on it, he would too. In early 1973, Akira reviewed my final draft, found my argument credible, and wrote to Professor Rattray saying so. And so it was that I became Dr. Ebbesmeyer.

That same year, Akira visited us in Dallas, where he was presenting a paper before the Entomological Society of America. We invited him to dinner and Susie fixed a favorite dish, Chicken Rice Roger, with heaping rice and chicken breasts. Susie, Lisa, Wendy, and I each had a helping, but before we could get seconds, Akira had wolfed down all the rest. We stared in ghastly fascination at this frail Japanese gentleman who could pack away food like a sumo wrestler. I marveled again when Akira and I and several other oceanographers gathered over crayfish at Deanie's, a restaurant by Louisiana's Lake Pontchartrain famous for its enormous dinners. The rest of us couldn't finish the piles of crayfish before us. While the beer and shoptalk flowed, Akira quietly disassembled all his crawdads with his tiny fingers (fingers that also folded exquisite origami dinosaurs for Wendy and Lisa), then finished all our leftovers. His stomach bulged from his rail-thin frame like a pig passing through a python.

This was the first hint I had of the ordeals Akira had endured as a child, and of the role that oceanic drifters, sown by the U.S. Navy, played in them. He never liked to talk about why he was so ravenous, but eventually the story came out. Akira grew up in Tokyo. He was about twenty when the war ended with the dropping of atomic bombs on Hiroshima and Nagasaki. Those explosions awed and alarmed the entire world, but most Japanese were affected much more by the privation that came before. The United States had tried an ancient strategy: starving its enemy into submission. By the spring of 1945, American aircraft, mines, and submarines had sunk more than 80 percent of Japan's merchant-ship tonnage. To eliminate the rest, from April to August B-29 bombers dropped more than forty thousand mines in the entrances to Japan's ports. The Allies, fully cognizant of the consequences, called their campaign Operation Starvation.

The mines quietly strangled Japan. When the A-bombs finally fell, the population was on the verge of widespread starvation. "Had the U.S. mine campaign started three months earlier," wrote Admiral Tamura Hiroaki, who commanded Japan's anti-mine defenses, "Japan would probably have capitulated before August. It would have been unnecessary to inaugurate the nuclear age at Hiroshima!"

Akira Okubo was one of the millions of Japanese who nearly perished during Operation Starvation. Hunger scarred his legs and left his hair copper colored. It scarred him emotionally as well; for the rest of his life, he ate as though he might never see food again. His appetite became legendary, even as his scientific stature grew. And he remained passionately curious about drifting mines.

The world shuddered at the radioactive fallout from Hiroshima and Nagasaki. Few know what was revealed on a single-sheet document, secreted by itself in an army-green steel box among the effects I inherited from Cliff Barnes half a century later: that thirty-five thousand live submarine mines placed around Japan to repel an American invasion were dislodged by two outsized typhoons in 1945, before minesweepers could recover them. Like landmines, these deadly drifters can remain dangerous for years, even decades, after they're forgotten. The last known live mines drifted into the Hawaiian Islands ten years after Japan surrendered. Fishermen from California to Alaska still occasionally snag mines that have sunk to the seafloor. Marine growth may cover a mine so fully that scuba divers can't see it from a few feet away. And it's still wise when beachcombing on remote Alaskan beaches to watch out for the horned mines hidden under piles of driftwood, ready to detonate with a nudge from a shifting log.

Akira and I both washed up in Seattle in 1974. He arrived to begin a promised professorship at the University of Washington, but somehow fell into a limbo between three departments—Applied Physics, Oceanography, and the Center for Quantitative Science, any of which could have used his talents. The university reneged on its offer and left him stranded and penniless, on the verge of starvation once again. Luckily, I could use his statistical skills in my sewage-tracking project for Metro. I paid Akira $2,000 to develop the mathematics to describe the dispersion of our drogues so as to incorporate the effects of hypersnarks in sewage dispersion, and it kept him going for six months while the university departments dithered over whether to hire him. Then he received an offer from the Marine Sciences Research Center at the State University of New York at Stony Brook, where he went on to work for the next twenty years. The method we'd meanwhile devised for calculating dispersion from drifters came to be widely used in oceanography.

Even after he landed at Stony Brook, Akira came to Seattle often—fifty times in twenty years, usually to perform contracts for UW, where he worked closely with Jim Anderson at the Center for Quantitative Science. Sometimes he would stay for a month and I would find an apartment for him. We would pal around, pursuing whatever drifty things sparked our curiosity. Our work was unsanctioned by academic or government entities and unconstrained by grant and contract requirements, but it led to the publication of six papers in peer-reviewed journals. We scoured UW's East Asian Library, finding and translating documents on transpacific Japanese drifters and their role in the nineteenth-century opening of Japan. We pondered the possible role of floating pumice in the origin of life. And we investigated the swarming behavior of insects and icebergs.

Akira was a consummate scholar and scientist, with little attention to spare for the mundane matters that occupy most people's attention. He was married once, but it did not last. He drove an old VW bus for a while but gave even that up, preferring to walk or ride a bike. He was utterly devoted to his studies.

Those studies ranged across a breathtaking range of subjects. What seemed like casual, disparate diversions yielded surprising insights and important findings. And these often looped back to illuminate the floating world. For example, Akira was fascinated with the author Edgar Allan Poe; while at Johns Hopkins, he visited Poe's Baltimore haunts. Often he would proclaim that Poe was an unrecognized oceanographic sage, as evinced especially by his stories "MS. Found in a Bottle" and "The Maelstrom." Later I would discover how much Poe influenced the evolution of oceanography.

Squirrels were another of Akira's improbable passions. When he wasn't teaching, we'd often meet outside the Boisserie café at the Burke Museum of Natural History on UW's leafy campus and sit beside an oak tree where we could feed the squirrels. One day, after we got our coffee and sat down under the oak, I noticed that all of Akira's tiny fingertips were bandaged. "What happened?" I asked. He explained that he'd run out of peanuts but his squirrel buddies gathered anyway. He tried to pet them and, in lieu of peanuts, they bit his fingers.

Akira's squirrel play disguised serious long-term research. Through much of the 1980s he studied the competition between native red squirrels and the introduced grey squirrels that had in one century overrun Britain, squeezing out the reds. This culminated in a classic paper, "On the Spatial Spread of the Grey Squirrel in Britain," published by the Royal Society of London. In 1980, Akira published many such accounts of ecological processes in a masterful book, *Diffusion and Ecological Problems: Mathematical Models.* Everything in nature was fair game for his mathematic virtuosity. Anyone could describe a process to him—icebergs and drogues for me, midges for the entomologist H. C. Chiang—and within a few days receive a five-page outpouring of calculations and formulae. Often it would be ready for publication, so carefully had he thought the problem through.

In the mid-1970s, Akira's work and mine made an improbable convergence. He'd been working on the kinematics of swarming midges with Professor Chiang. I'd joined him in that research while continuing to work on the icebergs that calve off Greenland and drift south to perish on the Grand Banks, like elephants trudging across the desert to their fabled graveyard. One thing led to another, and we began comparing these icy swarms to everything from elephant and wildebeest herds to the life cycles of stars. Finally, we found that mosquitoes offered a useful analogue to the swarming behavior of icebergs. Akira was delighted to have some new mathematics to work out. And we published papers on iceberg and mosquito swarms in two peer-reviewed journals.

It was the sort of inquiry that a conventional engineering approach would squelch—which is why I would get frustrated with consulting work, and eventually leave it behind.

In 1976 and 1977, as my work for Metro wound down, I resumed knocking on doors looking for work—this time at the Seattle regional office of the National Oceanographic and Atmospheric Administration (NOAA) and at the University of Washington. These efforts paid off, and I got to work on two interesting, instructive, and important projects. Only trouble was the first one nearly got me killed. Or rather, I nearly got myself and my intrepid pilot killed.

At the time, oil spills were an increasingly worrisome prospect around Puget Sound. A political debate raged over how much threat they presented and whether industry and government were doing enough to prevent them. NOAA was striving to figure out where surface currents would carry any oil spilled in the Strait of Juan de Fuca, the gateway to the Sound. But no devices had yet been devised to simulate spilled oil, so I set about inventing one that NOAA would fund. The problem was that the strait was too vast to survey with a vessel at the typical speed of ten miles per hour. I hit on the idea of using airplanes. I would lay sheets of thin plastic, six feet square, onto the sea surface. Each would have a large number painted on it, which could be spotted from a low-flying Cessna. Better yet, I would drop the sheets from the plane, saving more time and expense. It sounded simple, but it took some doing to perfect the scheme.

In those pre-GPS days oceanographers used the Mini-Ranger, a microwave tracking system from Motorola, to gauge position. I mounted one in the Cessna and read off positions as the plane flew low over a drift-sheet target. At eighty miles an hour, we could cover the strait eight times faster than a boat.

Next I had to test launching the sheets from an aircraft. I taped Venetian blind slats to the back of a plastic sheet, rolled it up, and tied a string around it, joined by an Alka-Seltzer tablet that would crumble in sea water and release the tie. Phil Taylor, a masterful marine technician at UW, flew the Cessna up to a few thousand feet and I shoved the roll out my door. Down it went, then sprang open as planned.

With the system evidently all worked out, I approached NOAA, which awarded me a contract to follow drift sheets in the western Strait of Juan de Fuca. In July 1977, I made about a hundred rolled-up drift sheets, rented the Mini-Ranger, and hired Phil to do the flying. We set up the two ranging stations on the shore and took off from Port Angeles on the eastern Strait of Juan de Fuca. When we reached position, I opened my door and ejected the roll. Phil had warned me to thrust it down hard so it wouldn't catch on the tail section. But however hard I pushed, it still got caught, sending us into a tailspin four hundred feet above the water. The stall alarm came on. Phil, a superb pilot, pumped the pedals. The plane gyrated to the left and right so violently that the drift sheet finally shook loose. We were just fifty feet above the water.

We had barely escaped ditching into the strait fifty miles from the nearest Coast Guard station, but Phil was unfazed. "Let's try that again," he said. I thought he was joking, but up we went. I tossed the second drift sheet, with the same terrifying result and another narrow escape. Now even Phil was convinced my scheme was not going to work.

When you need help at sea, ask the Coast Guard. I'd gotten along well with the Coasties who attended Cliff's classes, each of us helping the other with our master's projects. So I went to their station at Port Angeles and asked if they could launch my drift sheets. Sure enough, they were happy to help with oil-spill studies. They loaded all the drift sheets aboard their thirty-seven-foot cutter and spread them in a line across the strait. The rest of the project went as planned: The drifters zigzagged east and west but drifted toward the south shore. We'd shown that oil spilled here would be pushed by the winds sideways across the western strait and wash up on the U.S. rather than Canadian shore.

In 1976, I also hung around the Old Oceanography Building at the University of Washington, renewing old connections and forging new ones. One professor, Bruce Taft, knew my student work with water slabs at Dabob Bay. Now he was playing a leading role in POLYMODE ("Mid-Ocean Dynamics Experiment"), a joint U.S.-Soviet project to study eddies in the deep ocean, for which Congress had provided $10 million. I never learned what political purposes underlay this effort. Both nations' militaries must have been interested in the eddies, which like slabs can deflect sonar and conceal submarines. Whatever the project's motives, Taft asked me to join it.

This was a godsend; deepwater eddies were a vast, unstudied field, and ever since my student days I'd longed to track snarks in the open ocean. But doing so required enormous resources, far beyond those available to a graduate student. Now we commanded the resources of two superpowers, both eager to show their scientific stuff. No longer would I have to chase the elusive slabs in a surplus patrol boat. The U.S. Navy had recently requisitioned a new 182-foot ship, the research vessel *Gyre,* tailored for multidisciplinary oceanographic research, with a whopping eight-thousand-mile cruising range. I could live the dream as chief scientist aboard a state-of-the-art research vessel.

Round after round of planning meetings ensued, at Harvard and Woods Hole. At one, I proposed a pilot cruise to the Sargasso Sea to break in the team and test the advanced technology we would use in the main experiments scheduled for 1978. This flew, and I spent three weeks in April 1977 transecting the Sargasso Sea. This was my first deep-ocean work, and though it was only a pilot study, it was a revelation. I found ten snarks distributed through the sixteen-thousand-foot water column just south of Bermuda. Using historical data files, I traced these ten to their parent waters, scattered across thousands of miles of ocean, each of which had unique temperature, salinity, and oxygen signatures and a distinctive level of the

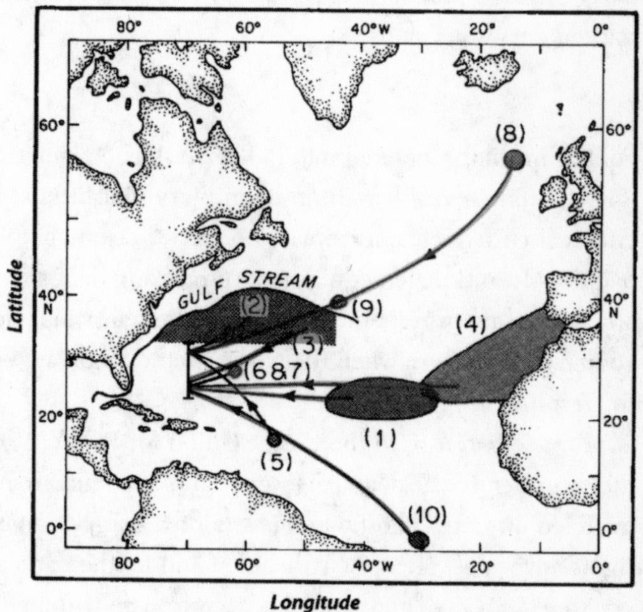

Ten snarks south of Bermuda tracked during POLYMODE, at depths of 600 to 17,056 feet. Ebbesmeyer traced these to home waters 212 to 2,808 miles away.

hydrogen isotope tritium. Like Dabob snarks, they retained the flavors of their origins. And like the coastal waters I'd studied, the open ocean was layered with discrete chunks of water. Dabob Bay was indeed a microcosm—for a sea of snarks.

In May 1978, we embarked on POLYMODE's most intensive phase: two vessels sampling in close coordination for two months, longer than I'd ever spent working continuously at sea. I took a twelve-hour shift, from midnight to noon, lowering a package of electronic sensors to two thousand meters (sixty-five hundred feet) and painstakingly transcribing the temperatures they recorded, just as I had at Dabob. This time, however, I plotted them on horizontal rather than vertical maps, to show the positions of the particular slabs we were chasing—snark hunters on the open sea. We found thirty-one snarks in a small area. Based on this and the 1977 samples, I estimated that ten thousand or more snarks inhabit the upper two thousand meters of the North Atlantic and many times that number roam the oceans worldwide. Each deepwater cast took about three hours, leaving lots of time to sip coffee and observe what floated on the Sargasso Sea. I was astounded at all the Styrofoam cups amid the sargassum weed, but I never guessed that someday I would track jetsam like that just as I now tracked deepwater slabs.

Politics inevitably entered into POLYMODE. At that time, so much of the Soviet Union's wealth went to its military that little remained for civilian efforts such as oceanography. The Soviet oceanographers in the project had grossly antiquated equipment, producing data so inaccurate that usually I could not use them. But to keep the political peace, we had to put a good face on things; when they used their own data, we made no fuss over the results.

I made two trips to the Soviet Union for POLYMODE, to Moscow in the summer of 1978 and Moscow and Georgia in April 1980. Politics intruded on both. Shortly before the first, two prominent Russian human rights advocates, physicist Yuri Orlov and mathematician Anatoly Sharansky, were convicted and sentenced to prison on trumped-up charges of treason. Several of us in POLYMODE sent a letter of protest to the Soviet Academy of Sciences, but I was the only signer going to Russia. In Moscow, I noticed a fellow at the bar keeping an eye on me. I went up and asked him how things were going. "Why did you send such a letter?" he asked. I replied that it was common in the United States for citizens to send letters of protest to congressmen. He realized that I was just a harmless benighted scientist and let up after that. He also asked how I could be a private oceanographer when all the other oceanographers worked for government or universities. I tried to explain what it meant to be an entrepreneur, but he didn't get it.

On the second POLYMODE trip our visas still had not been approved when we left for London. I had heard stories of other oceanographers who'd had to turn around when their visas did not materialize. Luckily, someone appeared at Heathrow and handed us ours. In Moscow, soldiers wearing great coats and carrying automatic weapons stood above and below every subway escalator and at the entrance to the hotel. Although the hotel was supposedly first-class, it had little or no toilet paper; everyone knew to bring his own. The tap water was an ugly brown, barely fit for a shower. The only place to get drinking water was in a restaurant—and even then I was suspicious. Better to drink beer or the ubiquitous vodka.

The trip coincided with our fifteenth wedding anniversary, and Susie and I had arranged to meet for an April-in-Paris vacation right after the POLYMODE sessions. Our delegation had a few meetings in Moscow, followed by a jet flight south to a Georgian town near Mount Elbrus, high in the Caucasus Mountains. Georgian officials greeted us and shared their best caviar and vodka. No one seemed to be shadowing me this time. I gave my talk on POLYMODE and water slabs, pausing for the translator after each sentence. It was slow going, but it went well.

Afterward, passing through the Moscow airport, I was shunted into a tiny cell surrounded by heavy bars; a steel gate clanged shut behind me. A soldier who spoke no English sat there, a red phone before him. He looked at my passport, said something in Russian, said something else over the red phone, and left. Another soldier with many more medals entered, looked at my passport, spoke into the phone, and left.

Finally what looked like a general, wearing one of those outsized caps reserved for officers of the highest rank, entered, examined my passport and my visa from the Soviet Academy of Sciences, and waved me on. The outer gate clanged open. As soon as our jet was airborne the passengers burst out clapping, celebrating our return to the West.

In December 1979, I made one more POLYMODE-related trip overseas, to present papers on the results of my North Atlantic snark chase to the International Union of Geodesy and Geophysics meeting in Canberra, Australia. As usual, I detoured on the way home. At that time, Quantas Airways allowed one free stop-off on its flights from Sydney to San Francisco, and I opted for a week in Fiji. Upon landing, I happened to pass a tourist shop selling one-week excursions to "a perfect holiday destination" called Beachcomber Island.

"Island" proved an overstatement. It was an atoll of one acre at most; I strolled its entire shoreline in twenty minutes. But I managed to get a room and soon fell in with a pair of Australian schoolteachers. We snorkeled endlessly around the calm, shallow reef, and I taught the teachers the trick of towing an inner tube full of beer.

I visited Australia again in 1985 to deliver a paper on the peculiar North Atlantic snarks known as "eighteen-degree water" (because they commonly form at that temperature) and to visit a POLYMODE colleague. This time I brought my daughter, Wendy, and stopped off for another week in Fiji and a one-day excursion to Beachcomber Island. Even then the island's name did not resonate with me.

3. Messages in Bottles

I am afraid that the picture the primitive peoples had of the sea was a truer one than ours.

—Thor Heyerdahl, *The Kon-Tiki Expedition*

Signatures of all things I am here to read, seaspawn and seawrack.

—James Joyce, *Ulysses*

As the 1970s faded into the 1980s, an upheaval in my parents' lives made them regular presences in mine—and, through their uncanny advice, led to a change in my life's mission. In 1977, at the age of sixty-two, my father was diagnosed with late-stage melanoma, a legacy of his chocolate-selling days. Fair-complexioned and prone to burn, he'd worn short-sleeved shirts and rested his arm on the window as he drove up and down the coast pitching Merckens chocolate. He'd long since left that business and had two other careers, as an engineer designing high-tension electric towers and an illustrator at UCLA's medical school. But that past exposure came back to hurt him. He underwent surgery to remove a large chunk of his upper left arm and received gold-injection treatments at UCLA.

More illnesses followed; from 1977 on, my father was never a well man. Arthritis kept him pretty much housebound, and he drank ample quantities of wine to dull the pain. He had congestive heart failure, and his systolic blood pressure soared as high as 310—deforming a valve and leaving him with just 25 percent of normal heart function. Doctors discovered cancer in his colon but were afraid to operate because of his weak heart. Lupus and Parkinson's disease weakened him, and he slowly went blind from glaucoma. But he fought on for nineteen more years, willing his body to function despite the progressive Parkinson's symptoms. Each morning he would step naked before a full-length mirror and command every cell in his body to stand at attention and get well.

Eventually, however, maintaining the half-acre yard in Van Nuys on which I'd grown up got to be too much. My parents realized they had to move to smaller quarters. I'd always told them we could do many things together if they'd come to Seattle. During one of their visits I pulled out a map of the city, scribed a one-mile circle around our home, and said they should buy theirs within it. In 1980, they did—just five blocks north of us.

We did find plenty to do. It was like the 1950s again, when Dad and I undertook endless projects. I had lunch with them several times a week—usually a Denver sandwich or creamed eggs on toast—and did chores around their yard, just as I had then. Mom became my clipping service, scanning the newspapers for anything about ocean currents. Dad drew the illustrations for many of my project reports and journal articles; working as I did without institutional support, I could not have completed them without him. Finally the shaking in his right arm got so bad he had to stop.

With funds left from the sale of their California home, my parents bought a yellow 1979 Chevy Caprice from a salesman who'd put just a few thousand high-speed miles on it. Mom named it Yellow Bird. When she passed away, she willed it to me, and I drive Yellow Bird to this day.

By this time, the business had grown and I had hired a number of people. Oceanographers love to party, especially this oceanographer, and at Evans-Hamilton West almost any occasion was an excuse to celebrate. Often Dad barbecued for all of us (one specialty: salmon on grape leaves from the vines in his new backyard). Despite all his illnesses, he always found a way to be happy. He even sought a floating refuge to match Lake Gregory, a reservoir high above the Valley where we retreated when I was growing up. (Dad bought a membership in a vacation community on Lake Gregory, complete with an eighth-acre lot, for $144. Motors were forbidden there, so we could row and fish in perfect peace.)

Curt and his father, Paul Ebbesmeyer, in 1986 with a thirty-inch Dolly Varden trout caught at Ross Lake by Curt's daughter, Wendy.

Dad found Seattle's equivalent to Lake Gregory at the Ross Lake Resort in the craggy North Cascades. "Resort" was putting a fine name on this collection of floating cabins, launched in 1951 when workers on Ross Dam, Seattle's hydroelectric source, lashed two of their bunkhouses onto cedar logs. Once again we'd found our own little floating world. I loved to watch the large water striders that paraded across the quiet waters between the cabins. Later I would study the pelagic striders, called *Halobates*, that walk the high seas, masters of their own two-dimensional universe.

Eventually, as Parkinson's overtook Dad, we flew him to Ross Lake in a float plane. But he got ten good years there. Meanwhile, Cliff Barnes, who'd been a second father to me, was falling under the shadow of Alzheimer's, though his amazing physical stamina still carried him along. Cliff often came to our Evans-Hamilton parties; I think it buoyed him to see his old students doing practical oceanography together. But as the years progressed, the disease stole his recognition of friends and colleagues. He no longer went to the office the university provided gratis with his emeritus status. Finally, Cliff's family asked me to clear out the office. I reconstructed it in our basement, in hopes he would stop by and it would rekindle some memories. Alas, this never happened. But as I picked through his papers I found he'd left me hundreds of notes on slips of paper marking the articles in journals and pages in books he wanted me to read—leads that would prove crucial guides to the floating world.

In the early 1980s, Metro planned two more big sewage plants in addition to the West Point plant whose outfall I'd studied—one at Renton, just southeast of Seattle, the other the Brightwater Treatment Plant to the north. This time I was contracted to gather data to determine where the Renton plant's outfall should be placed. In 1984 I began a series twenty-four one-day cruises for Metro, sampling temperature and salinity in minute detail all along Puget Sound's Main Basin. The agency had just purchased a new digital measuring system similar to the STD I'd used on Dabob Bay, but with the unfortunate acronym changed to CTD, for "conductivity, temperature, depth." (Water's conductivity reflects its salinity.) Once again I had to make a recalcitrant, untested measuring system work. After weeks of testing and tinkering, we lowered the CTD at a dozen sites spaced along Puget Sound, from Port Townsend to the turbulent Tacoma Narrows.

Again we found speeding hypersnarks—fast-moving water slabs formed by tidal collisions at the Tacoma Narrows—infesting the entire basin. And we found a staircase of abruptly differentiated water blocks, like a stack of variously colored poker chips, all interlayered. At the immediate, practical level, Metro's discharge scheme had to accommodate these stacked slabs, which trapped sewage and slowed its diffusion. On the global level, such temperature staircases and the barriers they present to mixing proved to be common throughout the world ocean. I would later find that the winds off Yucatán generated similar stacked snarks in the Gulf of Mexico, and that these broke down and intermingled off Louisiana and Texas.

Our Renton team also included Jack Word, an out-of-the-box thinker and innovator I'd hired to head our sediment-analysis laboratory at Evans-Hamilton. Contrary to the fond assumptions of Metro officials, Jack suspected that sewage carried a substantial load of oils and greases that were discharged from engine leaks, cooking, dishwashing, and all the other things people do with oil. These absorb all manner of heavy metals and nasty chemical compounds. He knew that some of them survived treatment and rose to the sea surface, but no studies had been done to confirm these observations. So Jack built a hydraulic test chamber and I helped him obtain actual sewage effluent and inject it into the chamber. A substantial portion rose to the surface. Jack and I also walked the beaches seeking the same effect in the real world. We spotted seagulls converging on the sewage outfalls and wondered if they were drawn by floating fats. One calm morning we rowed up slowly and quietly and got close enough to see them pecking up grease balls. After they flew off we gathered some of the balls as evidence.

Metro seemed to ignore these findings, though it accepted our other results. This was not a subject the agency was prepared to deal with. Sanitary engineers are very slow to change their thinking and practices; they tend to do so only when federal agencies such as the EPA order them to. The EPA itself strenuously resisted, but we prevailed. Our findings stood up so well to scientific scrutiny that they were eventually published in a peer-reviewed book.

Running Evans-Hamilton's western region had its perks, above and beyond Dad's barbecued salmon. Each year the board of directors—Doug Evans, Bob Hamilton, and I, with our wives—held its annual meeting in the sort of marine location befitting an oceanographic enterprise: Cancún, Barbados, the Cayman Islands, St. Maarten. Then we discovered just how vulnerable our business was to a reverse oil shock. In 1986, the OPEC countries gave up trying to set production quotas, and oil prices plummeted to a now-mythical-sounding $9 a barrel. In a single month we lost all our petrowork—about three-quarters of our income—as the panicked oil companies canceled work orders. The Caribbean trips came to a crashing halt. We faced the prospect of bankruptcy because we had ignored one of the most basic rules of business: stay diversified. I'd broadened our client base in the West, but the Houston headquarters remained overwhelmingly dependent on oil. Doug, Bob, and I had gotten our first jobs in the oil patch, and we'd never had to leave it—till now.

Then another prospect beckoned. The federal Minerals Management Service (MMS) oversees the drilling industry and its environmental impacts. We heard through the oceanographic grapevine that the MMS would be taking bids for a $1 million study of frontal eddies off North Carolina, the cascading turbulence produced as warm water spins off the sharp drop in temperature—or front—along the inshore and offshore edges of the Gulf Stream. (The MMS and state of North Carolina worried that oil spilled offshore could be whirled onshore by these eddies.) Better yet, the job was set aside for a small business such as ours. If we won it we could stay in business, but our cash-flow problem would remain. We would not receive any payment until a year after the oil crash, and we needed $400,000 to keep going.

Banks were no help, but salvation appeared from an unexpected quarter: Bob's wife, Glenda, who had received an ample inheritance. In 1932, her father was working as a soda jerk in Chattanooga. A customer, impressed at the way he handled the counter, offered him a 10 percent share in an innovative new chain of restaurants if he'd manage them. It was a risky move, leaving a steady job during the Depression, but it paid off. The result was Krystal, the second-oldest hamburger chain in the United States and the source of the nest egg that kept us in business. We dodged ruin and built Evans-Hamilton into a company with thirty employees and an annual income of $5 million.

Determinate drifters—floating objects released to track currents—had come to dominate my work. In 1987, for the MMS's frontal-eddy studies, I deployed satellite-tracked drifters for the first time. We had to be extremely judicious with them because it cost so much to receive satellite data back then—a monthly cable bill times ten for each buoy—and once released they could not be recovered. And so I acquired the data for only a three-week portion of our study. We tracked several frontal eddies in real time—much as I'd tracked open-oceans snarks in POLYMODE but with more difficulty, because some eddies would start forming and fizzle out. We were like surfers trying to gauge which wave to paddle after, with limited aircraft and ship time for chasing and just eight precious satellite buoys.

We managed to catch two eddies that maintained their identities and named them after the comedians Eddie Murphy and Eddie Harrington (who performed as Bud Abbott in Abbott and Costello). We had set up an array of current meters so that when the eddy passed we could measure the currents it spun off. I expected the eddies we studied to get swept along with the Gulf Stream and hold together all the way across the North Atlantic, if they could clear Cape Hatteras without getting squished. Eventually they would hit Europe and drop off the debris they'd swept up along the East Coast.

It would have been wonderful to follow these watery tumbleweeds. I would have seen how they trap flotsam, from spilled oil to Styrofoam cups, and transport it across the Atlantic to Europe—counterparts to the meddies that form off the Strait of Gibraltar and travel nearly intact to the American coast. Such trapping can prevent oceanic mixing, a significant impact. But our resources and reach were limited; we were only funded to work off the North Carolina coast. Every environmental study I've worked on has left me with a greater sense of the ocean's big picture—and a greater residue of frustration. Each time I took a few steps toward understanding that larger picture, I was stopped short by the study's terms.

Years later, however, I got a teasing snatch of evidence of the frontal eddies' transatlantic movements. The late Nick Darke was a celebrated playwright and dedicated "wrecker," as beachcombers are traditionally called in Cornwall, at the southwest corner of England, where he lived; his beach faces right into the incoming Gulf Stream. He showed me two pieces of parquet flooring that washed up a few feet apart there. They fit together perfectly, and Nick surmised that they'd crossed the Atlantic together. If so, they'd been held tight in a small eddy, perhaps a frontal eddy.

That same year I deployed drogues in Prudhoe Bay, off Alaska's North Slope, as part of a study to gauge the effects of a five-mile gravel causeway being built through the shallow water to Endicott Island, the first continuously operating offshore oil field in the Arctic. To mitigate the impact on sea life and let fish pass, it was proposed that the causeway be breached with culverts and bridges. But that raised another question: Would water rush through these gaps or flow around them and disrupt coastal flow and inshore habitats?

In the end I did not get to make that call. As in so much consulting work, I was a bit actor in a larger drama, contributing studies that others would use to make their determinations. But I concluded that the causeway itself substantially disturbed the coastal flow; it would have been much cheaper to build bridges from the start, as was eventually done. It's the same story whenever we disturb coastal processes: We always pay a stiffer price if we don't do the right thing in the first place.

On March 5, 1986, I conducted another drifter study—one of my favorites—closer to home. A large resort on Port Ludlow, a small bay near the entrance to Puget Sound, sought a permit to discharge more sewage into the inner bay. Local residents opposed the permit, and the resort's owner asked me to mediate with them. At issue was how fast the bay flushed—how fast new water replaced the old. Trouble was, flushing is difficult to measure. It usually entails taking temperature, salinity, and current measurements and inserting these into complex mathematical models that tend to mystify juries and other laypeople. I had a budget of just $10,000, hardly enough to deploy and recover expensive current meters. So I decided to try something new: to measure the flushing directly.

To do this, I proposed placing three hundred drift sticks—one-by-four wood slats weighted and held upright in the water with flotation collars, which I fabricated using strips cut from foam sleeping mats and duct tape. Atop the sticks I set red and yellow flags, flying about three feet above the surface. I spread them throughout the bay and waited to see how long it took to flush them out. Developer and residents alike would see clearly how fast the bay purged itself. I had no expectations; I was prepared to stay in the field for a week or more. To track the drift sticks, I sped around in a Boston Whaler and, using a sextant, marked the position of each stick on a hydrographic chart every few hours—during daylight. I did not have the funds to place lights on the sticks and cheap GPS trackers were not available back then.

The winds were calm; this was fun in the sun. Knowing that even a light wind could influence the result, I rose several times during the night to double check; the water remained glassy smooth, and my boat's flag drooped around its pole. At daybreak I discovered to my amazement that all but four of the three hundred drift sticks had exited the bay, and those four were hung up in the brush along the shore. At first I thought someone had taken them. A sailor from the resort marina volunteered to take me far and wide to see where the sticks had gone. A day of searching showed that swift tidal currents had dispersed them across a reach of eleven miles. It was the fastest flushing I had ever seen. So clear and compelling was the result that both the developer and the anxious residents were satisfied. Environmental studies rarely have such happy endings.

Necessity had made me turn to drift sticks at Port Ludlow, but they proved an unbeatable research tool. In using them, and in following drift markers generally, I was plugging into a hundred-and-fifty-year-old oceanographic tradition—or a thousand-year-old one, depending on how you want to count.

The first people we can be reasonably certain made use of determinate drifters were the Norsemen who colonized Iceland in the ninth and tenth centuries. I first learned of their proto-oceanographic ventures from a 1962 doctoral dissertation by Iceland's Unnsteinn Stefánsson, which I discovered—bound in sharkskin and inscribed to Cliff Barnes—when I sorted out Cliff's office after he was incapacitated by Alzheimer's disease. In the true Cliff Barnes tradition, Stefánsson had thought to look where no one else had—in the Norse historic sagas—and found detailed observations of the movements of objects on the waters around Iceland. I went on to chart the drifts of seven such objects, plus one coffin, described in a twelfth-century genealogical record, *The Icelandic Book of Settlements.*

The god Thor told the Norsemen where they should locate their settlements: As they approached the island, they should cast overboard their most cherished floatable objects, their ceremonial bench boards and high seat posts, and settle wherever these washed up. One ailing chieftain, Old Kveldúlfur, felt his life slipping away as his longboat approached the new

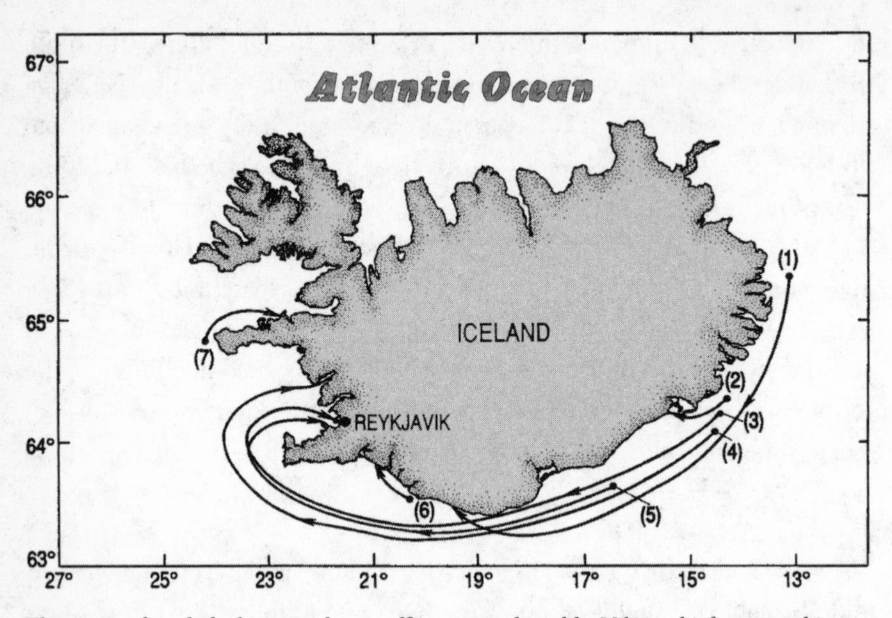

The routes of six drift objects and one coffin cast overboard by Viking chiefs approaching Iceland in 860–930 CE, *as described in the Icelandic Book of Settlements: (1) Lodmund's high seat posts, (2) Hrollaug's seat posts, (3) Thord Skeggi's seat posts, (4) Kveldúlfur's coffin, (5) Ingolf Arnarson's seat posts, (6) Hastein's bench boards, (7) Thorolf Mostur-Beard's seat posts.*

land. He directed his son Egil to cast his body overboard and make a home wherever it landed. And so Old Kveldefur continued to lead his clan even after he became flotsam.

Like Egil Kveldúlfurson, Ingolf Arnarson fled Norway to escape retribution for a killing. And like any good Viking, he threw his seat posts overboard as he approached Iceland. He named the well-favored harbor where they washed up Reykjavik, "Smoky Bay," after the volcanic vents nearby. More settlers followed as their posts and bench boards washed up there—and thus Iceland's capital came to be.

Thor's instructions actually represented sound practical oceanography. They led the colonizers to the sites where whales and driftwood—precious on nearly treeless Iceland—would collect. This was the first use of determinate drift markers that I know of.

Two centuries later an exiled Japanese poet also cast out drifters, hoping they would find their way to his old home and perhaps bring him back there. Akira first told me this story, recorded in the medieval epic *The Tale of the Heike*, and said he believed it to be factual. In 1177, in the fallout from a palace intrigue, the poet Yasuyori was banished to a remote shore. He wrote a thousand poems lamenting his plight on the small wooden planks called *stupa* and cast them to the waves, hoping—or, if he knew the local currents, expecting—that some would reach his parents 155 miles away. One stupa washed up near the palace and was taken to the emperor, who, moved by Yasuyori's words, recalled him from exile.

Yasuyori's stupa were messages in bottles minus the bottles. Seven centuries later a fad for sending bottled messages swept the world, nudged along by a writer, Edgar Allan Poe, better known for claustrophobic tales of terror than freewheeling transoceanic communication. Again it was Akira, with his devotion to Poe, who first suggested this idea. Akira always admired the oceanography in Poe's stories, particularly in "MS. Found in a Bottle," published in 1833 to great acclaim. This tale of a doomed mariner, the last survivor on a ship, drifting toward an inexorable fate, spoke powerfully to an age obsessed with long-distance communication, from clipper ships and the telegraph to the soon-to-emerge penny post. And it seems to have exerted a powerful influence on the way people view and use the ocean—with a boost from Charles Dickens, who published his own tale of a fateful bottlegram, "A Message from the Sea," in 1860.

I realized all this when I uncovered three obscure but brilliant papers published in 1843 and 1852 in the *Nautical Magazine and Naval Chronicle* by its editor, Rear Admiral Alexander Becher. Becher's magazine was the journal of record for "bottle papers," as messages in bottles (MIBs for short) were called then. In that era, with its hunger for good data about ocean currents, these drifters were deemed to have important scientific value, and they were still rare enough to be widely publicized.

Becher described the 174 bottles known to have been released and recovered from 1808 to 1852 in the North Atlantic Subtropical Gyre, which we call the Columbus Gyre after the first mariner to master it. From Becher's accounts I constructed a timeline of the number of MIBs released in each of those years. It showed several interesting historical correlations. No bottles at all were reported released from 1813 to 1816, perhaps because of the Napoleonic Wars and the War of 1812. The number of releases began to increase in 1819, the year of the first transatlantic steamship voyage. And it more than tripled after Poe's story was published, to an average of eight per year—a tally that has continued to rise in the century and a half since.

Becher and Poe were both farsighted. Becher was the first to chronicle the passage of drifters around a gyre, using the same sort of monitoring system—a network of beachcomber informants—that I would revive in the 1990s. Poe launched a fad that endures into the Internet age; he may even have transformed fishing as well as oceanography. When Poe popularized the idea of sending messages on the water, he broadcast the surprising fact that glass vessels could survive the open ocean and pounding surf and land safely on shore. A few years later, the fishing industry seemed to take the hint. Until then, the size of fishermen's nets—and hence their catches—had been limited by the heavy, awkward, water-absorbing wooden floats used to suspend them. Then, in the early 1840s, Christopher Faye revolutionized fishing with the glass fishing float. Buoyed by Faye's floats, nets became larger, dramatically increasing catches. For better and (considering today's grievous overfishing) worse, the Industrial Revolution had reached the fishing industry.

Students of the sea, likewise realizing the potential of nearly unsinkable glass, began launching their own MIBs. And so Poe unwittingly incubated the emergent science of oceanography. The American, German, and British navies, eager to understand the currents on which they sailed, began routinely launching research MIBs. From 1885 to 1887, Prince Albert I of Monaco, an avid oceanographer, released 1,675 bottles and other drifters from his schooner *Hirondelle*, mostly in the mid-North Atlantic.

Durable though glass bottles had proven to be, another venturesome drift experimenter insisted on even sturdier vessels. Rear Admiral George W. Melville, who had survived an epic ride on an Arctic ice floe to become the U.S. Navy's engineer in chief, had fifty streamlined, football-shaped, metal-tipped wooden casks released from Point Barrow between 1899 and

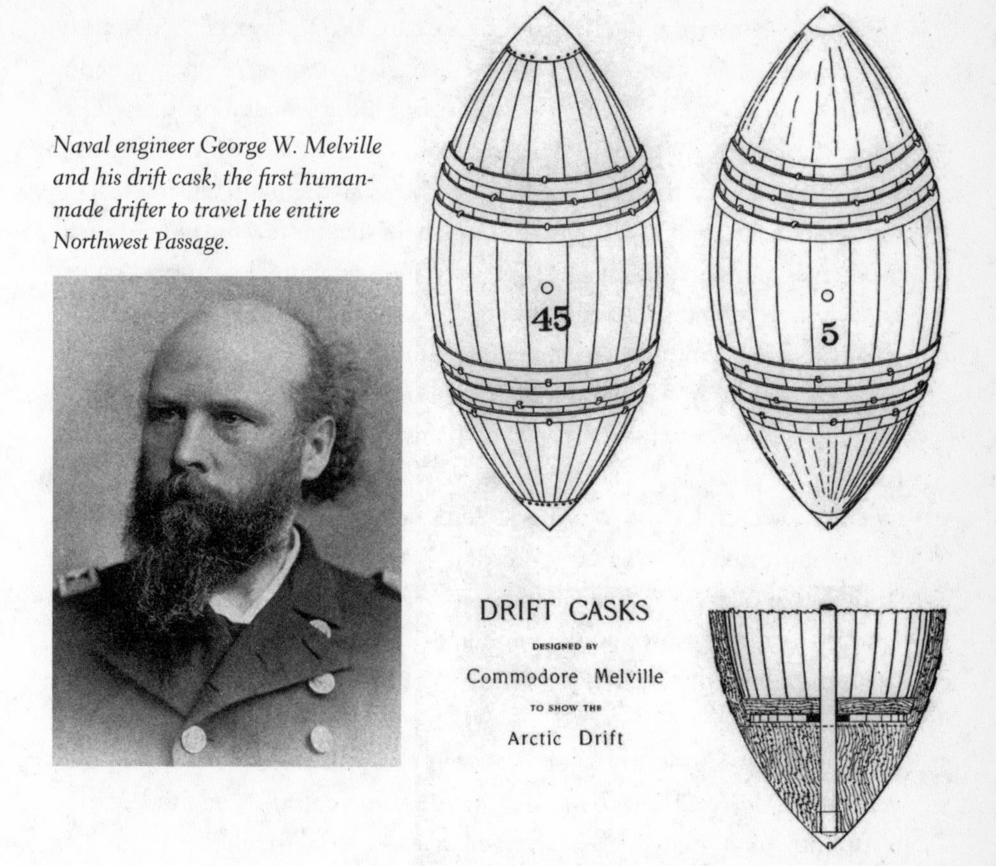

Naval engineer George W. Melville and his drift cask, the first human-made drifter to travel the entire Northwest Passage.

1901. One reached Siberia in 1902, another Iceland in 1905, and a third Norway in 1908. They were the first human-made drifters to travel the entire Northwest Passage.

Many beachcombers dream of finding just one message bottle. Wim Kruiswijk has found more than six hundred; he has an astonishing knack for spotting washed-up objects that other eyes, including mine, have missed. Kruiswijk, who once operated the Beachcomber Museum in the North Sea town of Zandvoort, has compiled a sociology of message bottles—the first attempt to analyze statistically what inspires people to cast their words onto the sea. More than a third of the 435 MIBs from twenty-nine countries he tallied were from people seeking pen pals. A quarter gave only their senders' addresses, leaving finders to guess their motives. Nearly

13 percent came from one Belgian fisherman, asking finders to send postcards from their home towns. Eight percent shared jokes. Six percent contained religious messages, from a "get a free Bible" operation in Evreux, France, a preacher in Lowestock, England, six girls on a Mormon cruise, and an ex-boxer in Kent who proclaimed he "saw the light." A small share were released in school projects. Others held love notes, drawings, advertisements, pornography, or pleas for help. One contained a protest against pollution, which evidently did not include bottles tossed at sea.

Others have found even more ingenious reasons to cast their words upon the waters, from propaganda to personal expression. After the Dutch painter Henk Noorlander finishes with his preparatory sketches, he rolls them up in wine bottles and tosses them into the North Sea as gifts to far-off, unknown collectors. During its long hostilities with Beijing, the Taiwanese government released more than one hundred thousand canisters stuffed with flags, propaganda, and consumer goods—soap, perfume, wash towels—that were scarce on the mainland. Taiwanese activists also released thousands of bottled leaflets protesting the 1979 sentencing of the dissident Wei Jingsheng.

Frivolous or profound, such messages could float for many years, even centuries. Message bottles are like mobile time capsules, left to the eyes of unknown strangers. They are sacrificial scouts, like the space probes we launch into the galactic ocean knowing we'll never see them again, though we hope to hear from them from places we cannot reach ourselves. Space probes function as calling cards for our planet and species; MIBs introduce us to people we will almost surely never meet. The difference, of course, is that MIBs cannot report back themselves; someone else must find them and reply. But they operate on the same timescale as the earliest space probes, which have been sending back word for about thirty years.

Perhaps that fact helped induce a venerable brewery to commemorate its two-hundredth birthday with message bottles. I've heard from beachcombers scattered from Washington to Florida and Texas who noticed something different about certain beer bottles they spotted in the jetsam. Beneath the sand and barnacles, these bottles were embossed with a

Guinness's ocean promotion: one of two hundred thousand commemorative message bottles and the proclamation from King Neptune furled inside.

harp, a world map, and a message: "1759–1959, Special Bottle Drop (Atlantic Ocean) to celebrate and commemorate Guinness Bicentenary, 1959." Inside, scrolls and an official seal showed dimly through the brown glass.

It wasn't until BBC Radio interviewed me for a program titled *Message in a Bottle* that I learned the full story of how these bottles got there. The stunt was the brainchild of Guinness's formidable chairman, A. W. Fawcett—A.W.F. to his employees, who joked that the initials stood for "Appointment With Fear." In 1954, five years before the bicentenary, A.W.F. cajoled his shipping cronies into dropping fifty thousand of the commemorative bottles at eleven sites worldwide. When the big date rolled around, he had thirty-eight vessels bombard the Atlantic and Caribbean with a hundred fifty thousand bottles.

A.W.F. thought of everything; his brewery even took out insurance against beachcombers cutting themselves on broken bottles. The scrolls inside conveyed advice from King Neptune on how to recycle the bottles into decorative table lamps. Each of the scarcer 1954 bottles also included a little metal leprechaun, with a card proclaiming that it had kissed the Blarney Stone. (The company actually arranged to have each case of a thousand or so leprechauns dangled over the wall at Blarney Castle until it touched the stone.) But the big challenge was to seal the bottles well enough to last at sea. A.W.F. settled on a triple seal—pill cork, metal cap, and lead wrapper—and predicted it would last "at least five hundred years." It was, he added, "perhaps intriguing to stop and think that many of these bottles, floating in the oceans of the world, may survive even many of the people reading this particular memorandum."

"Officially, the story was that this was to aid research into the perfect sealing of the bottles," Guinness representative Sue Garland told the BBC. Then she admitted the real motive. "Lots of people around the world would find the bottles, write in and make stories for the newspapers."

Regardless of motive, the seals worked. Even now, Guinness receives two or three letters a year from people who have found bottles. One came from an inmate on Isla Santa Maria, a penal island seventy-five miles off San Blas, Mexico. He described the discovery in almost biblical tones. He was "stone-breaking at the edge of the beach" when he "saw floating on the water a bottle which contained a parchment" with an enticing message. Another informant reported that Inuit hunters had found about eighty bottles washed ashore near Coats Island, at the mouth of Hudson Bay. They used them for stone-throwing practice until they learned that Guinness offered a reward.

Years after the great bottle drop, the British Admiralty requested Guinness's drift records.

Sometimes the sands on shore conspire with currents offshore to conceal message bottles for many decades. Then rising seas and storms—indicators of climate change or decades-long weather cycles—sweep away the sands and expose the bottles. This effect seems especially pronounced in Alaska, the source of the most epic accounts of long-lost bottles showing up.

In 1994, National Weather Service employee Jack Endicott found a bottle on a beach near Yakutat, still in excellent condition. Its printed message asked the finder to report it to the International Fisheries Commission in

Seattle for a twenty-five-cent reward. That might seem miserly today, but the deadline given was March 31, 1935, when twenty-five cents was real money.

Even that six-decade burial and resurrection pales against the saga of the "czar's bottle," which culminated at Strawberry Channel Beach near Cordova, Alaska. During an interlude between storms in the winter of 1994–95, Brooke and Gayle Adkinson were searching for fresh flotsam when they spotted an amber-colored bottle poking out of a wave-cut dune. It bore no special markings, but they saw a waterlogged paper inside, teased out the pieces with tweezers, and wrote to ask my advice. I instructed them to try to assemble the pieces, a jigsaw of texts printed in Japanese, Russian, and English. That made it possible to decipher the essential information: "Vladivostok . . . East Siberia . . . The Pacific Hydrographical Expedition," and, on the other side, "Imperial Russian . . . thrown 5/18 July 1913, N 54°26', E 141°55'." Those coordinates indicate the bottle was thrown overboard in the Sea of Okhotsk, just north of Sakhalin Island. But why was the czar's navy conducting research in this bleak, empty sea a year before the start of the First World War?

The answer lies in the fierce Russo-Japanese War of 1904–5, the first time in modern history that Asian forces defeated a European power. Russia lost in part because its ships had to sail around Africa to reach the front in Manchuria, and the Baltic Fleet arrived too late for the decisive battle of Tsushima Strait. Afterward, looking toward the future, the Imperial Russian Navy began exploring Siberia's ice-choked waters, seeking a shorter route to the Far East. It built two icebreakers, the *Taymyr* and the *Vaygach,* which set out north from Vladivostok in July 1913. They worked their way along the fogbound inside passage between mainland Asia and Sakhalin's western shore, avoiding the strong southward current that sweeps down the island's east side, and headed toward the Sea of Okhotsk and Kamchatka Peninsula. When the *Taymyr* rounded the Sakhalin's north cape and entered the sea, its crew tossed out message bottles, including the one that Gayle and Brooke would find eighty-two years later.

How did that bottle travel from Sakhalin to Cordova? It got an assist first from the same strong southward current the ships had avoided, which sped it along at about fifteen miles a day. By year's end it had traveled far enough south to escape getting caught in the ice that covers most of the Sea of Okhotsk. Somewhere south of the Kurile Islands, probably off the Japanese island of Hokkaido, it entered the North Pacific Subtropical (a. k.a. Turtle) Gyre. OSCURS showed that two to three years later it arrived off British Columbia's Queen Charlotte Islands and turned north toward Prince William Sound, where the waves buried it in the sands sometime around 1917—just as revolution swept the nation that launched it. And there it slept through Armistice Day, the Russian Civil War, another world war, the cold war, and the Korean and Vietnam conflicts. Another of the czar's bottles recently showed up on Spitsbergen, and I'm now tracing its launch site and pathway. No doubt even older bottles still float on the currents or hide in deserted dunes, patient messengers from another age.

It's hardly surprising that preachers and proselytizers should also turn to message bottles in their campaigns to spread the Word. The Bible is seeded with marine and aquatic wisdom. "Cast thy bread upon the waters, for thou shalt find it after many days," urges Ecclesiastes. "The Lord is upon many waters," notes Psalm 29. "Thou tellest my wanderings: put thou my tears into thy bottle," advises Psalm 56. "Blessed are ye that sow beside all waters, that send forth thither," declares the prophet Isaiah. And, of course, "except a man be born of water and of the Spirit, he cannot enter into the kingdom of God."

I don't belong to a particular religion; organized religion just gets in the way of too many things. But it's hard to escape the spiritual side of the ocean, and of oceanography. Such talk makes most oceanographers and other scientists uncomfortable; I know all too well the blank, discouraging stares my colleagues give if I don't purge any religious allusions from my speech—as well as the stiffening body language of religious groups when I speak to them about the ocean and they come to suspect I'm not born-again.

Nevertheless, religion and the sea are deeply entwined, even in ocean science. In the mid-1800s, Matthew Fontaine Maury, the superintendent of the U.S. Naval Observatory and oft-hailed "father of oceanography," inaugurated modern ocean science after a burst of biblical inspiration. Laid low by illness, Maury asked his son to read from the Bible. He was struck by a reference in one psalm to "the fishes in the sea and the pathways they are in" and vowed that if there are pathways in the sea, he would find them. The result was his pioneering 1855 volume, *The Physical Geography of the Sea*, as close to a bible as a science can have.

"People, especially people in crisis, are naturally attracted to water," writes Malidoma Patrice Somé, a healer from landlocked Burkina Faso. Jesus recognized that fact. It's no accident that he gave an awful lot of sermons on the beach and chose most of his disciples from the ranks of fishermen: "He that would learn to pray, let him go to the sea." Likewise he who would find followers capable of deep devotion, and evangelizers seeking souls susceptible to bottled messages.

Sometimes these messages land with uncanny accuracy. In 1983 and 1984, while trekking Baja California's three thousand miles of arid coastline, the adventurer Graham Mackintosh encountered first temptation, then timely spiritual aid inside beached bottles. He had reached Malarrimo Beach, a desolate, scythe-shaped protrusion into the Pacific four hundred miles south of the California-Mexico border. Malarrimo, as I can attest from experience, is a beachcomber's paradise; debris from all around the North Pacific washes up there—sometimes even the makings of a well-stocked bar. Mackintosh found intact cans of beer and bottles of rum, vermouth, brandy, scotch, Japanese whiskey, and London gin. He couldn't resist; he loaded up and, half-pickled, struggled for days through the soft sands. Finally he stumbled on another washed-up bottle and found a religious tract entitled "Help From Above," decrying the evils of strong drink. He discarded the booze and got on with his trek, for which he later won the prestigious Adventurous Traveler of the Year award.

I discovered the granddaddy of all bottle evangelists by a process only slightly less serendipitous—or, if you will, destined. It started in a hospital waiting room. Inside, my father was recovering from surgery to remove two large tumors from his colon. A story mentioning my work had just appeared in the *National Enquirer*, and Dad was having fun telling the nurses about his son the "tabloid oceanographer." (They replied that clearly that son had an inquiring mind.) I sat outside with my mother, brother, and wife anxiously awaiting the surgeon. Suddenly a scrap of paper sticking out of the Gideon Bible on the lamp table caught my eye. Scrawled on it was a note directing readers to Psalm 56:8. "Thou tellest my wanderings: put thou my tears into thy bottle: are they not in thy book?" To me that meant I should turn my attention to tract-filled bottles.

Thereafter, as I pored through old books and periodicals in the University of Washington's Suzzallo Library, amid the comforting scent of browning paper, I came across curious reports of seagoing bottled messages. The popular magazines of the 1940s and 50s—the *Saturday Evening Post*, *Coronet*, and *American Magazine*—carried occasional features about a Reverend George Phillips of Tacoma, the first to make the sea his pulpit. I wondered if Phillips was still there, and if he had any useful information on sea currents. But for all his past fame, not a single acquaintance, colleague, or local church I asked had ever heard of him. Comparing city directories, I found a Phillips still listed at one Tacoma address since the 1940s and made a call. I discovered that Reverend George had died but his widow, Ella, aged ninety, would be glad to tell me how she and her husband launched forty thousand sermon bottles.

The story actually began during the Depression, when a former silent screen actor, decorated World War I combat officer, and recovering alcoholic named Carleton E. Null began preaching the gospel along America's roadways. As they drove across the country, Null and a companion would hurl tracts rolled in brightly colored cellophane at roadside tramps. These drive-by missiles, which looked like firecrackers, became known as "gospel bombs." Null boasted that his 10 million bombs yielded fifty thousand new churchgoers.

When Null reached Tacoma on his gospel-bombing barnstorm tour, he provided half the inspiration George Phillips needed to start preaching. The other half came from the sea. "I was down on the beach one day in April 1940, not far from our home," Phillips later recalled, "and I saw the tide carrying drift wood. Why couldn't I spread the gospel in the same way?"

Phillips, like Carleton Null, was a recovering alcoholic, and he relished the idea of bottling a different sort of spirit—of sending salvation in the vessels that once afflicted him. "Whiskey once got me down," he declared. "Now I'd like to see it bring men up." He abandoned his previous work, selling real estate and used cars, and devoted himself full-time to his World Wide Missionary Effort. Though it might seem primitive, his crusade actually rested on two twentieth-century innovations: mass literacy and mass-produced, machine-made bottles, which debuted in 1903. He became a pioneer recycler, scouring streets, alleys, garbage cans, and city dumps for usable bottles.

George and Ella Phillips's forty thousand gospel torpedoes landed up and down the Pacific Coast and in Mexico, Hawaii, New Guinea, and Australia. Two even washed up, unbroken, among the wrecked homes and bodies when Hurricane Audrey ravaged Louisiana in 1957. They elicited fifteen hundred replies, a thousand of them pledges to stop drinking. Hundreds of respondents vowed to return to church.

One former Chicago businessman found a Phillips bottle while beachcombing with his young son near Acapulco, where he'd fled after his firm went bust and he became estranged from his wife. The sermonette inside ended with the warning, "Be sure your sins will find you out." Jolted by such pointed advice, he returned home, rebuilt his business, paid off his creditors, and reconciled with his wife. And, he wrote to Phillips, "I owe it all to that bottle that came out of the sea."

More evidence of George Phillips's transoceanic impact showed up near Yakutat, Alaska, the day after a tsunami alert didn't pan out. Two beachcombers, Dawn and Tony Laforest, found an old bottle containing an unusually personal evangelical pamphlet by a Captain Walter E. Bindt. Captain Bindt described himself as a Hawaiian missionary's great-grandson who'd gone to sea in 1922. "For over twenty-five years I have been plying the ocean lanes of the Pacific," he recounted. "In 1947 the Lord brought Brother George Phillips of Tacoma, Washington, and myself together. On learning of the remarkable blessing from distributing the Gospel in Tide Bottles, I realized immediately that I was to have a part in this ministry. Since then, it has been my privilege to toss many thousands of bottled Gospel Bombs in the oceans, sometimes 1,500 on a single voyage. Returns have come from many foreign shores and from the United States."

Others heard the call; inspired by the many articles on Phillips, bottle churches sprang up round the world. Proclaiming that "the oceans are His

Glass floaters that survived at sea including, from the left, a "gospel bomb" from the Gospel Afloat campaign in Cloverdale, California; a Japanese fishing float; message bottle released in the North Pacific by Captain Basil Biggs from the MV Bonnieway; a light bulb; message in a cognac bottle released from Oahu.

pulpit," the Merseyside Bottle Evangelists of Liverpool launched 65,000 bottles and received 5,500 replies. Ethel Tinkham, the eighty-two-year-old widow of a Baptist minister in Portland, Oregon, packed tiny tracts into 3,950 baby food bottles. Some reached Japan and New Jersey. A textile-mill supply clerk named Jewel T. Pierce launched 27,800 scripture-packed booze bottles down Alabama's Coosa River. Some of them crossed the Atlantic, and at least one entered the Mediterranean.

All told, Reverend Phillips and his fellow bottle bombers cast about three hundred thousand scriptural MIBs onto the waves in the mid-twentieth century. Since then the practice has largely died out—superseded, perhaps, by televangelism and other more efficient preaching media. But oracular messages still wash up now and then. In 2001, twenty-one-year-old Mary Raikes picked up a battered plastic pop bottle on the shore near her home in Harrington, Maine. It contained just a cryptic, anonymous message: "If you need help, go north."

"At the time," Raikes told me, "I probably felt like I needed help. I figure not many people get these, so I decided I should follow it. Alaska was north, so I went." She found a summer job as a state park ranger in Ketchikan. As they say about books, does the person find the bottle or does the bottle find the person?

How many message bottles have been cast on the waves? Crunching the three hundred thousand gospel torpedoes I've tallied and the share of Wim Kruiswijk's bottles that contained religious tracts, I'd estimate the total number of MIBs released since the mid-1900s at 6 million—five hundred thousand of them from oceanographers.

But what's the fate of messages injected into the floating world? How many of them actually find their readers? We can only gauge that from return rates, the number of people who find bottles and other drifters who write in response or mail back reply cards. (If that sounds like the language of pollsters, that's because casting MIBs is a type of polling.) I've tallied the rates for thirty-two oceanographic, evangelical, and promotional campaigns totaling about 1 million bottles and cards. Return rates vary widely, from a high of 50 percent for oceanographic drift cards released along the densely populated North Sea and Puget Sound to as few as one per hundred launched around uninhabited Antarctica. Overall, my best guess is that for each ten bottles released one is actually found, opened, and reported; three are found but go unopened or unreported; three wash up on remote shores and are never found; one gets buried in beach sands; one drifts for a decade or longer; and one suffers any of a number of other fates—smashed on rocks, swallowed by an animal, sunk by a leaky seal or the weight of barnacles.

Some turn up much later, in unlikely circumstances. A beachcomber brought a 1950s Scripps bottle to a lecture I gave in the Westport (Washington) Museum. A bottle released by the midcentury oceanographer Dean Bumpus surfaced at an estate auction in Tennessee. While searching for artifacts from the Manila galleon *San Felipe*, lost near Malarrimo Beach, my companions and I found ten message bottles. Clayton Krause, my guide to beachcombing Malarrimo, says his Mexican friends have found hundreds in the sand dunes.

Different messages and different containers elicit different responses. A dull request receives far fewer replies than one that offers the finder good wishes, spiritual uplift, or hard cash. Gospel bottles get relatively high responses, but even these vary widely. Jewel Pierce reported a 20 percent response rate on the Coosa River, a sign perhaps of the level of piety in Alabama. Another gospel bomber, Everett Bachelder, achieved an impressive 10 percent return rate with bottles released in the remote Bering Strait. George Phillips got a 7.5 percent response promising an "interesting and attractive Booklet—Free." A promise of "spiritual aid" brought a 4 percent response from across the Pacific in the Philippines to the ninety thousand bottles Reverend Kraig Rice launched from California.

On the scientific side, Prince Albert I elicited an astonishing 14 percent return rate, even though beachcombers were far less numerous in the 1880s than they are today. He hedged his bets by using strong glass bottles, ballasted for minimal wind exposure, with a message printed in nine languages asking the finder to report his discovery. What beachcomber could resist answering a royal summons?

The Seamen's Church Institute of New York City achieved reporting rates as high as 12 percent by dangling prizes that included a fully rigged miniature ship in a pint bottle, a three-foot model of the famed clipper *Flying Cloud*, and painted seascapes. Californian beachcomber Alan Schwartz launched 250 bottles and achieved an impressive 6 percent response from the Philippines, Japan, Thailand, and Vietnam. Schwartz told me that in addition to using stout bottles and indestructible seals, he "enclosed two copies of my message so the finder keeps a souvenir after a copy is returned. I wish a very superstitious part of the world good luck, first thing in the message. I put an uncirculated U.S. $1 bill in every bottle. U.S. currency is the universal language and pays for the return reply up front. All in all, I believe in money. One wonders at the return rate if I'd used $5 bills."

Despite such ingenious individual projects, a vast store of oceanographic data assembled through a century of drift-bottle releases has been lost to history. In 1976, the U.S. Navy decided for political reasons to move its oceanographic office from Washington, DC, to Mississippi. In the course of that move, it apparently destroyed reports on hundreds of thousands of bottles launched from Maury's day up through the 1950s. In an era of computer analysis and electronic measurement, these laboriously compiled records were considered antiquated and outdated.

Having gathered and employed both kinds of data, I knew how very wrong that assessment was. For years I despaired over the loss of the bottle records. Later I found a way to replace them, by reprising on a larger scale what Admiral Becher and his beachcomber network did in the early nineteenth century.

As the 1980s drew to a close, I began to get the message myself—from the currents, if not in a bottle. The good old days of Mother Mobil were gone; the oil crash had taken the fun out of consulting oceanography. Engineers seemed to throw up more and more headwind on environmental projects. I came to suspect I could find better ways to spend my time than banging my head against brick walls.

In 1990 I decided to cut my consulting work (and my salary) by 40 percent. Consulting had allowed my curiosity greater play than Mobil or a government job would have, but following flotsam had freed it to sail on even wider seas. I'd learned that the floating world held many other kinds of drifters besides drogues, drift cards, and sewage effluent, and I wanted to trace more of them. I resolved that I would take no funding or advertising and see where my curiosity led me. As it turned out, the decision to cut my hours was propitious; soon I would be immersed in another pursuit that has held me to this day.

I intended to write a book on what Cliff Barnes and I had discovered about Puget Sound, with the working title *The Sound of Water Bodies.* But this plan did not hold for long. Even as I plotted my semiretirement, fate and the stormy ocean were making other plans—and a container full of sneakers was tottering atop a freighter in the middle of the North Pacific.

4. Eureka, a Sneaker!

The ocean is forever asking questions and
writing them aloud on the shore.

—Edwin Arlington Robinson, *Roman Bartholow*

The year 1990 was a watershed for me, and for flotsam science. No sooner had I decided to cut back on consulting work and concentrate on my own research than the other shoe dropped—literally.

On May 27, 1990, the cargo vessel *Hansa Carrier*, en route from Korea to Los Angeles, hit a severe, sudden storm. Stowing cargo is critical on today's ships, whose decks are packed up to seventy feet high with eight-by-ten-by-forty-foot steel shipping containers. Distribute the weight unevenly and the ship may heel over in a storm. Lash it too loosely and she may shed containers like a dog shaking off fleas. Cargo practices have since improved, but in the 1990s as many as ten thousand containers may have gone overboard each year. The largest known spill, during a 1998 typhoon, dropped three to four hundred containers into the mid-Pacific.

The *Hansa Carrier* was named after the Hanseatic League, a medieval German merchant alliance with a famous seafaring history. But the ship's reputation was less illustrious; her crew was notorious for stowing containers sloppily, so much so that some of the shippers who'd used her called her

My son, the oceanographer: Curt's mother, Gene Ebbesmeyer, with two of thirty-four thousand Nike cross-trainers lost in a later North Pacific spill in 1999.

"the ship from hell." When the May 27 storm hit, she lost twenty-one containers. Five were crammed with Nike shoes—78,932 sneakers, hiking boots, and children's shoes. The pairs were unlaced, so each shoe became a separate piece of flotsam.

As usual, the shipping industry guarded the news closely; shippers and manufacturers tend to keep their cargo spills secret, to avoid embarrassment and liability. But then the shoes themselves blew the cover off. Eight months later, in January 1991, after drifting two thousand miles eastward, the Nikes began beaching on Vancouver Island. The prevailing winter winds and currents then pushed them north as far as the Queen Charlotte Islands. Then the winds shifted, as they do each spring, and blew south through the summer. Thousands of shoes stranded along the Oregon coast, just a few score miles from Nike's headquarters outside Portland.

The news media loved the story. One day, when I stopped by my parents' house for our usual lunch of creamed eggs on toast, my mother pulled out the newspaper story she'd clipped on the Oregon shoe strandings, and I said I'd look into it. The floating world had given me the call.

To the ancient Greeks, Nike was the winged goddess of victory, though it was another god, Hermes, who lent the hero Perseus the winged shoes that let him fly like a bird. In the late twentieth century, "Nikes" became first guided missiles, then sneakers. To me, they represented a chance to have a little fun with ocean currents, and a welcome respite from sewage, oil spills, and offshore platform designs.

Like an oceanic gumshoe, I set about tracking down the beached sneakers. When I ask shippers about flotsam washing up from their container spills, 95 percent stonewall. "We are not aware of any lost merchandise," goes the standard answer. I could never find out what was in the other sixteen containers that washed off the *Hansa Carrier* that day. But Nike's transportation department was refreshingly open. Its staff provided the date, latitude and longitude, and the container load plan, listing each container's contents down to the last shoe. Unlike many other footwear manufacturers, Nike stamps a simple identification number, a "purchase order ID" (POID), on each of its shoes. It tracks its products so meticulously that it can trace a single shoe to its mother container. For example, the POID on a three-year-old Nike recovered on Maui, 90 04 06 ST, indicated that the Korean factory (ST) had received the order in April (04) for delivery by June (06) 1990.

Following the POIDs would eventually enable us to answer the question I'm always asked when I speak on the Great Sneaker Spill: How many of the five Nike containers broke open in the storm and disgorged their contents? We've found hundreds of shoes from each of four containers and not one shoe from the fifth. Someday, somewhere on the Pacific seabed, submarine archeologists will find 17,112 sneakers nestled in a giant steel shoebox.

With the information Nike provided, we could determine conclusively whether any shoe that washed up had fallen off the *Hansa*. The first beachcombers' reports of sneakers washing up—"forerunners," as one reporter called them—were equally specific. I had something that's very rare with spontaneous flotsam (as opposed to determinate drift markers): both point A, when and where an object starts to drift, and point B, when and where it washes up.

With this data in hand, I thought of OSCURS, the Ocean Surface Current Simulator program that Jim Ingraham had developed at NOAA to calculate the effects of ocean currents on salmon migration. I wondered if OSCURS would work as well with inanimate drifters as it did with swimming fish; subtract the fish's swim speed and you should have flotsam. In the decades since grad school, Jim and I had gone our separate ways, but the sneaker spill rekindled our friendship. I called him, and he was glad to help.

I decided to stage a blind test of OSCURS. I gave Jim only point A, the *Hansa* spill, and asked if he could calculate point B, when and where the shoes would wash up. "I'll fax back the answer in an hour," he replied.

OSCURS homed in like a carrier pigeon, making direct hits on the earliest point Bs—November and December 1990 on the Washington coast and January and February 1991 on Vancouver Island—where the first sneakers washed up. Jim and I needed more data—the dates and locations of other wash-ups. We began seeking out beachcombers in Oregon and asking if they'd spotted any Nikes, but it was a slow, laborious process. Then we hit the jackpot: One beachcomber directed me to Steve McLeod, a painter in the easygoing resort town of Cannon Beach. Steve is a classic starving artist; he'd won some recognition and big commissions but refused to play the corporate game, preferring to follow his muse and eke out what living he could. He's also a dedicated beachcomber, which may nurture his muse and certainly boosted his living on this occasion.

Twenty years earlier, as if by premonition, Steve had painted a picture of two gigantic hiking boots hovering over an imaginary beach. When he started finding beached Nikes, a light went on. This was the heyday of sneaker chic, when newspapers feasted on stories about inner-city kids shooting each other for their Air Jordans. Scrape off the barnacles, toss the washed-up Nikes in the washer, add a little bleach, and they looked and felt like new. Steve became a mail-order matchmaker for hundreds of people up and down the coast who'd found mismatched sneakers and wanted mates. He negotiated swaps: a size 10 left for a size 9 right here, a right size 12 for a left size 7 there. Soon everyone wound up with matched pairs, and Steve wound up with a shoestore's worth. When I visited his loft in Cannon

Beach, it was filled with two-by-four racks covered with drying sneakers. He sold them on the street, along with his usual trinkets, for $30 each, and took in $1,300.

"Is the shoe worth its salt?" quipped Jim Ingraham. He and I both took to wearing sea sneakers—in my case, a pair of fluorescent-pink Nike Flights that Steve gave me.

Better yet, Steve collected even more data than he did sneakers. He had notes on where and when sixteen hundred shoes had washed up; without him, I might have collected only a third or half as many reports. All told, we located "point B" for 2.5 percent of the shoes washed off the *Hansa Carrier*—almost as good as the 2.8 percent reporting rate for thirty thousand scientific message bottles released near the same site in 1958 and 1959 as part of the International Geophysical Year. Media coverage, swap meets, and Steve's diligence had proven nearly as effective as bottled pleas at eliciting responses from finders. It was solid data, good science. And it was the beginning of a flotsam-monitoring network that now circles the globe—thousands of sharp-eyed field monitors, volunteers in the search for telltale flotsam and indicator jetsam.

The sneaker spill introduced me to the world of beachcombing, a culture I had only brushed up against before but from which I've since learned a great deal. Beachcombing appeals to deep-seated impulses and aspirations—to the scientist, explorer, collector, and treasure hunter in everyone and, deepest of all, to the inner hunter-gatherer. It is poor man's oceanography, research as play, unconstrained by professional ambition and open to everyone with eyes to see and feet to walk.

Many beachcombers are uncelebrated salt-of-the-earth types; some are salty dogs. If they weren't seeking washed-up treasures and curiosities, they might be home keeping scrapbooks or restoring old cars. Often they have little formal education, but they're far more intelligent and inquisitive than many of the academics I've known. Beachcombing is a hobby—an unfashionable word these days—that opens up on everything else. Franklin D. Roosevelt credited his uncanny knowledge of world geography and history to his lifelong passion for stamp collecting. Likewise, to be interested in beachcombing is to be interested in everything. Beachcombers are the keepers of the ocean's memory, sifting and sorting the chaotic surfeit thrown up by waves and tides, transmuting trash into artistic and scientific gold.

Even after drifting for three years in corrosive salt water, the shoes remained surprisingly supple and wearable; teenagers can do much more damage to them in one year than the sea did in three. Like seabirds, the sneakers easily rode out storms. I studied their floatability up close, in a hotel spa. Once they became saturated, the sneakers floated upside down, their soles even with the water surface; the high tops of my size 12 Flights cut the water below like the keels of sailing ships, their laces twirling like a jellyfish's tentacles. At sea, these tough soles would protect the softer fabric from biting birds and bleaching sun. And, we confirmed with each beached sneaker, nothing grew on the soles. Foiling barnacles is no mean trick. Perhaps Nike could supply antifouling coating for ships' bottoms.

The secret of the shoes' buoyancy was the microscopic gas-filled chambers that Nike inserted to provide springiness and absorb shocks. "Air Jordan" was a misnomer; this gas was actually sulfur hexafluoride (SF_6) gas. Because SF_6 is inert and detectable in vanishingly small concentrations, it's also proven valuable in oceanography; researchers would inoculate blobs of water with it and followed them across the sea. Only trouble is, it's an extremely potent greenhouse gas, with more than twenty thousand times the global-warming impact of an equal quantity of carbon dioxide. Nike has spent millions to find an alternative gas and eliminate SF_6 from its shoes, as part of a broader effort to make its products less environmentally burdensome.

Even today, I get a warm reception at Nike headquarters, where I'm called the "ocean doctor" and invited to speak now and then on shipping, flotsam, and the sea. That's not surprising, considering all the unexpected good publicity Nike reaped from the follow-up to the *Hansa Carrier* spill. Its shoes' durability at sea impressed everyone. "Ten Nike marketing managers could huddle for a week and not come up with a better metaphor for the company's mission—to get Nikes, with near-biblical image making and Yankee dispatch, into the hands of everyone on the planet," Bruce Grierson wrote in Canada's *Saturday Night* magazine seven years later. "It's tempting to think the whole thing was planned."

Over the years, many people have asked in all seriousness if I masterminded the Great Sneaker Spill. No, I respond. It's just the inevitable result of shipping billions of shoes across the sea each year. In fact, Nike has always refrained from exploiting this potential marketing bonanza. When asked why, one company manager replied, "Floating on the water is not a sports attribute we wish to endorse."

OSCURS proved so accurate at describing and forecasting the current-borne sneakers' movements that we published the results in *EOS*, the peer-reviewed weekly newsletter of the American Geophysical Union, which goes out to some thirty thousand scientists worldwide. That triggered a new tsunami of attention from both the popular media and the oceanographic

Drift of the great Nike sneaker spill of May 27, 1990, as simulated by OSCURS. N marks the spill site, the stippled plume indicates the drift path, and the dots at upper right show dates and locations where thirteen hundred shoes were discovered by beachcombers. At Station Papa, indicated by P, oceanographers released 33,869 message bottles during the International Geophysical Year.

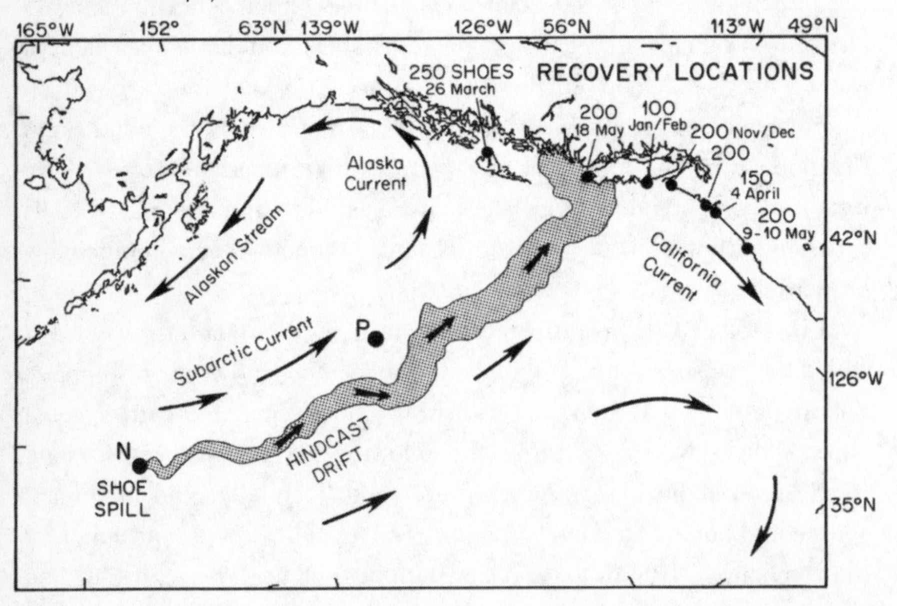

community. Everyone from the *National Enquirer* to National Public Radio ran with the floating sneakers. The headlines were soaked in puns: "Afoot on the Sea," "Footloose Sea," "Sole Travel," "Sole Survivors," "Lost Soles Searching the Currents," "Not a Sole Was Saved," "Soles Lost at Sea Sneak up on Beaches." "Running water" took on a whole new meaning. David Letterman announced his sixth-best reason to join the navy: "If you find a Nike sneaker floating on the water, it is yours." Even the checkers at my supermarket had seen the stories. They called me "the tennis-shoe oceanographer."

But my own colleague Jim Ingraham had the most fun. He titled a seminar we gave at NOAA's Monterey Bay lab "The Strange, the Bizarre, and the Science of All Things Afloat on the World's Oceans: An Introduction to Flotsammetrics of the North Pacific Ocean by Curtis Ebbesmeyer (the Strange) & Jim Ingraham (the Bizarre) at the National Marine Fisheries Service (the Science)."

More important, the sneaker study had a conspicuous and, I like to think, lasting effect on oceanographic education. At least half a dozen books recount it, and I still get letters from teachers asking me to help shape their syllabuses. It's deeply gratifying to think that some kids might dedicate their lives to studying and protecting the oceans after getting an inspirational nudge from the saga of the telltale sneakers.

The sneakers tested OSCURS's capabilities on a scale far beyond what Jim had intended. Until then he had only been able to check his model against the three-month trajectory of a satellite-tracked buoy. Now he could test it against thousands of drifters over an entire year. Come winter, the coastal winds and currents and the sneakers headed north again. The next year they turned back south, and by summer 1992 Nikes were washing up along the coast of northern California. After another year, the California Current transported some to Hawaii. Jim and I flew there to see for ourselves. Norman Shapiro, photo editor for the *Honolulu Advertiser,* presented us with a high-top hiking boot—its interior still velvety after three years adrift—he'd found in Kahana Bay on Maui's eastern shore. Other Nikes beached on northeast Lanai and the northern tip of the Big Island.

OSCURS next predicted that the sneakers would reach Japan in 1994.

The alert went out in prime time; I stood in my pink transpacific Nikes on Japan's NHK-TV and asked viewers to report any beached Nikes they found. But that doesn't seem to have been enough to mobilize informants; no Asian wash-ups were ever reported. We had to wait until the sneakers rounded the North Pacific Subtropical (a.k.a. Turtle) Gyre—a planetary vortex four times the size of the United States—and made landfall again on America to pick up their trail.

The year 1992 was a busy time. In my professional work, I analyzed the data from chasing frontal eddies off North Carolina and determined how much contaminated sediment the state ferries were stirring up along Seattle's waterfront. On my own, I tracked Nike sneakers down the West Coast. Then, on Labor Day, a fax appeared on my desk—a news clip reporting that a flotilla of toys was invading Sitka, Alaska.

The flotilla was actually a mini-menagerie: washed-up plastic turtles, frogs, beavers, and ducks. Thad Paulson, the editor of the *Daily Sitka Sentinel*, was as puzzled as everyone else. He ran classified ads seeking clues and was deluged with replies from readers who'd found them. Unlike the Nike sneakers, the toys had no serial numbers or manufacturing data. Only the ducks bore a clue: the words "The First Years" and a logo showing children at play, embossed on their little chests. At the *Sentinel*, intern reporter Eben Punderson and librarian Evelyn Bonner traced them to a single container that had been shipped from Hong Kong by the company Kiddie Products, Inc. on the Evergreen-line containership *Ever Laurel*, bound for Tacoma.

At two o'clock in the morning of January 10, 1992, a meteorological battering ram broadsided the *Ever Laurel*—what would be called a hurricane if it happened in the Caribbean. Large containerships receive periodic weather reports customized for their routes—every twelve hours in those days—but storms can come up much faster than that. This storm came up far too fast to be tracked. Its forty-foot waves took the *Ever Laurel*'s crew by surprise. They hung on as she heaved from side to side, listing thirty-five degrees or more. The wings of her bridge dipped into the waves.

Suddenly the containers nearest the pilot house burst their steel lashings. Containers usually fall in stacks; a ship rarely loses just one. This time twelve tumbled over the side. One held 28,800 bathtub toys. Once in the water, the toys had to complete five more escapes before they could float free. They were glued in sets of four to cardboard backing. These sets were enclosed in plastic housings. Three sets were packed into a cardboard box, and a dozen boxes were packed into a larger carton banded with a plastic strap. The cartons were encased in the heavy steel container.

The waves snapped the container's door latches and other barriers, the seawater pulped the various cardboard layers, and the turtles, ducks, frogs, and beavers escaped. Where did they go, and along what route? In order to ask OSCURS that question, we needed the date, latitude, and longitude of the spill. But that data resided only on the freighter's log sheets. Its owner, the shipping company Evergreen, refused to divulge them for fear the unflattering news would cost it customers. Evergreen finally granted us permission to interview the freighter's captain, on the condition that we promise not to disclose his, his vessel's, or the company's name. But we still had to wait for his ship to dock, then scramble on short notice when it did; so efficient has shipping become that crews unload hundreds of big containers in half a day.

Finally, after a year of waiting and cajoling, we met "Captain Mike," the *Ever Laurel*'s skipper, who was as genial and helpful as his employers had been recalcitrant. Our four-hour interview ranged over toys, shoes, container lashings, oceanography, and practical navigation. Captain Mike explained the fine balance a captain must steer between the threat of losing containers and the need to meet tight schedules amid North Pacific storms. Twice a day he computes the time needed to complete the voyage along two alternate routes. One, farther north, is shorter and, in good conditions, faster by several hours. But it's prone to worse storms and often takes longer because the vessel must slow down in high waves. The other route, to the south, is longer but typically faster. By gauging conditions and weighing the relative time lost to distance and waves, the captain determines the fastest course each day. After verifying his calculations several times, he compares his choice with routes plotted by shore-based oceanographers. In a

Curt and Jim Ingraham with tub toys recovered from the January 10, 1992, spill.

three-hundred-hour voyage, the choice of route typically makes a difference of just three hours, but that's still a significant contribution to the shipper's thin profit margin.

Captain Mike never said anything directly about the toy spill. But he left his log open to the date the containers went overboard. At last we knew: The toys went overboard at longitude 178.1 degrees east and latitude 44.7 degrees north—near the international date line, halfway between the equator and the North Pole at the junction of the Turtle and Aleut gyres. We could begin tracking them in earnest.

Plastic bath toys might at first glance seem inauspicious drift markers. They don't come with preaddressed return forms, as scientific message bottles do. They don't carry messages at all, or even Nike-style. And they're nowhere near as valuable as hundred-dollar sneakers.

Nevertheless, the bath toys inspired even more public enthusiasm than the shoes. That seems to reflect in part the peculiar popular fixation on so-called rubber duckies, which have migrated from infant bathtubs to adult fashion and design. The yellow ducky is an icon of whimsy, nostalgia, childhood innocence, and pop-cultural kitsch. It rolls the teddy bear, pink flamingo, American eagle, puppy in the window, and Mickey Mouse into one insistently cheerful cartoonish image.

Once we started soliciting reports of washed-up bath toys, we discovered that many species of ducky are afloat out there. According to the *Guinness Book of World Records*, one collector has amassed a mind-boggling nineteen hundred toy ducks. A firm called Great American Duck Races rents yellow ducks with sporty blue sunglasses for charity "races" worldwide. In this variation on both racing and roulette, spectators pay to "adopt" the numbered ducks, which are then dumped in a river, and the first to float past the finish mark wins its patron a prize. In 1994, two years after the spill, Great American supplied 1.5 million ducks to 127 races. Fifty thousand web-footed racers floated past Singapore's skyscrapers, and a hundred thousand "raced" down England's lazy River Avon. With such volumes, some inevitably escape the post-race cleanup crews and reach the open sea.

Fortunately, we could easily distinguish these and other "false canards," as Galveston Island beachcomber Cathy Yow called them, from our target drifters. The *Ever Laurel* critters, designed by the prominent child psychiatrist T. Berry Brazelton for The First Years, Inc., toy company are unique among all the toys adrift. The ducks, made of hard plastic rather than rubber, scarcely resemble the Disneyesque duckies of popular lore and decor. Their shape, like those of the other three critters, is angular and unexpressive, simplified rather than stylized, cubist rather than cartoonish.

I received wash-up reports for a full 3.3 percent of the toys spilled, a 30 percent higher rate than the sneakers elicited—even though media attention turned the orphaned mascots into collectibles, making finders loath to part with them. Years after the spill, many still adorn kayaks and fishing boats. Beachcombers tend to recall when they found the fabled tub toys the way they recall their first kiss. "A guy in a Pelican bar offered me $50 for the red beaver we'd found near Surge Bay, Alaska," English teacher Daniel Henry recounted, "but I couldn't bring myself to do it." Even other animals seemed susceptible to tub-toy mania. Henry's wife found their beaver "within an otter's midden pile of crushed clam and crab shell two hundred feet into the forest from the high tide mark . . . The otters' curiosity and love of play led us to believe that a young 'un might drag up a plastic beaver to a knoll in this ancient, moss-dripping forest. Looking for a friend, the pup simply chose the toy closest to his species."

Not surprisingly, many of the toys sported animal bites, just one of many abuses they had to endure at sea. The surf smashed them against the rocks, rupturing many. Under the bleaching sun the yellow ducks and red beavers turned albino, though the turtles and frogs retained their blue and green. I knew that some would likely drift north into the Arctic pack ice, so I performed some tests to see how they might survive there. I chilled sev-

The drift of twenty-nine thousand tub toys from spill site (T) as simulated by OSCURS, and the dates and locations where approximately four hundred toys were discovered by beachcombers. The toys passed the 1990 Nike spill site (N) and Ocean Weather Station Papa (P). The thick line shows the toys' estimated trajectory allowing for a boost in speed and slight deflection from the wind. Circles mark six-month intervals. OSCURS projected they would land near Sitka at the time of the first actual recoveries, November 16, 1992. The thin line shows the slower trajectory of objects unaffected by wind, with triangles marking six-month intervals and a large triangle for the point reached on November 16. Black squares mark reported toy recoveries and small symbols mark the points reached on January 1 and July 7 of 1992, 1993, and 1994.

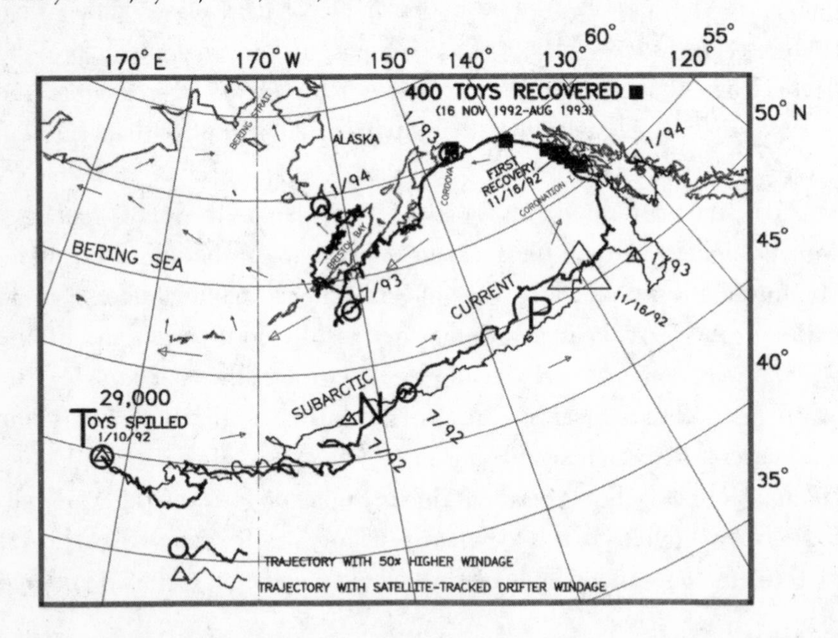

eral toys in my freezer, then tried to smash them with a hammer. They did not crack. I drilled holes in four and immersed them to see if they would still float filled with water. They did; the plastic itself was less dense than sea water.

My colleagues thought the toys would soon disintegrate into plastic confetti under the assaults of sun, sea, and ice. But they floated on, bleached and battered but still recognizable after sixteen years. I shouldn't have been surprised; I knew that plastic lobster-pot tags and drift cards had circled the North Atlantic for twenty and thirty-one years, respectively, with their inscriptions still readable. Later I learned from the manufacturer that the toys had been designed to survive fifty-two dishwasher cycles.

In spring and summer 1993, thousands of tub toys sped at five to eight miles a day past Alaska's glaciers and volcanoes. In September, Chrystle White, serving at Shemya Air Force Base in the outer Aleutian Islands, reported the westernmost wash-up of all, three hundred miles west of the international date line. She and her comrades gathered up more than two hundred of the tub toys and airlifted them to the Anchorage Salvation Army store. After that, OSCURS predicted that thousands would drift on past the Aleutians to Siberia and thousands more would loop south back to Sitka. If they survived—and it seemed they would—a few might circle back after three years to once again beguile beachcombers in Washington and British Columbia.

The toys' longevity yielded an oceanographic bonanza, but it demanded a far longer commitment than I'd intended; I did not expect to track tub toys for a quarter of my life. But the outcomes so intrigued Jim that he kept extending his OSCURS projections, far beyond anything he'd contemplated for scientific drifters. First he programmed OSCURS to cover the initial drift, from spill to first landfall, halfway across the Pacific. The results were so persuasive that we braved peer review and published an account in *EOS*, as a companion piece to our report on the sneaker spill. None of the scientists reviewing them had any negative comments.

Interesting container spills were now occurring every two years or so, but we hadn't heard much more about the toys. They were hard to spot on the beach and, we thought, would quickly disintegrate in the sun. After all, the sneakers had proved difficult to follow for more than a year or so; their serial numbers faded and it was hard to get beachcombers to scrape off the tongues, which were not so barnacle-proof as the soles, and read the data printed on them. We figured we would forget the toys and move on to the next spill.

We figured wrong. The toys kept their distinctive shapes. The media loved them; Jim and I simmered in a hot tub for eight hours while *People* shot endless pictures of us with ducks. Beachcombers, spurred by the attention, kept picking them up. So I typed out a flier seeking information on the "Most Wanted Duck," and Jim and I passed out copies along the Washington coast. It was the beginning of what would become the *Beachcombers' Alert* newsletter and network.

Jim kept at it for years, updating his drift animations month after month with U.S. Navy weather data till the day came when OSCURS predicted that the toys should strand on the Washington coast. On November 10, 1994, right on schedule and in another blind test, a fellow named Vern Krause spotted a faded yellow duck while scanning for glass floats on Pacific Beach. He ignored it, thinking a kid had dropped it. "You'd better pick it up," his wife Shirley urged. She remembered our flier.

More reports trickled in. Karen Gerber found a turtle, its beached belly covered with scratches, in the dune grass at Ocean Shores. Young Guthrie Schweers found two turtles and four frogs down the beach from the Langara Light House in the Queen Charlotte Islands. Jim and I were elated; OSCURS had forecast these recoveries within a month, excellent results considering the toys had drifted for thirty-four to forty-four months. Alas, I did not recognize at the time the importance of what these drifts suggested: the steady orbital period of the Aleut Gyre, the first step in understanding the periodic relationships of all the seas' gyres and the way they fit together like gears in a clock. The floating world had dropped a hint, but I would not take it for years to come—not till my eyes were opened at, of all places, a beachcombers' fair.

Meanwhile, I received valuable encouragement from kids at the schools I visited to talk about the sea and its flotsam. "In these rubber duck stories people could have hated your ideas and laughed at you each step of the way," Brandon Sherick-Odom, a student at Harbor Day School in Corona del Mar, wrote in a letter he sent me afterward. "This story of tracking them has taken years and is not complete yet, that takes patience." The kids called me Dr. Duck.

Once again, serendipity opened a door on the floating world. In her book *The Zuni Enigma: A Native American People's Possible Japanese Connection*, the Alaskan anthropologist Nancy Yaw Davis cited the toys as an example of the sort of transpacific drift she argues could have brought Japanese travelers to America some seven hundred years ago. In January 2002 she contacted me when she stopped in Seattle for a book signing and broached the idea of a scientific meeting on Pacific currents and winds and their effects on human mobility. I suggested we tack on a beachcombers' fair, something Alaska did not yet have.

We both thought Sitka, Alaska's capital when it was still Russian America, would be the perfect venue. Nancy was born and raised in Sitka. It was there that Bishop Ivan Veniaminov, Russia's answer to Ben Franklin, had pursued his wide-ranging studies in ethnology and natural science. And, of course, it was there that the beached bath toys had first been discovered.

"Maybe next year," Nancy said.

"Why not this year?" I asked.

"But there's no time to raise funds."

"We can do it without funds," I countered. Writing grants seemed a waste of time that could be better spent beachcombing—and any senior scientist who really wanted to come would find a way to get there. And so it was that six months later, in July 2002, Nancy convened Pathways Across the Pacific and I held Sitka's first beachcombers' fair. Betty Meggers from the Smithsonian Institution, Ben Finney of the University of Hawaii, Jim Ingraham representing NOAA, Nancy, and I met at the Eddystone Inn, overlooking the water. A local fisherman named Larry Calvin took us out beachcombing aboard his salmon boat, *Morning Mist*. Betty, Nancy, Jim, and I returned the next year for Pathways II, accompanied by several members of the local Tlingit people, who mastered these waters long before Russians or white Americans ever saw them.

In 2004, at Pathways III, the gifted father-and-son beachcombing team of Dean and Tyler Orbison displayed a hamper filled with 111 tub toys they'd found scouring the same area near Sitka for the previous eleven years. Dean had presciently recorded the date and location of each finding. I quickly calculated how many toys they'd found each year and saw that recoveries had peaked in 1992, 1994, 1998, 2001, 2004—on average, every three years, allowing for the usual variation in natural processes. This confirmed a cycle that was evident in the bottles released from Ocean Weather Station Papa. The same cycle would also be observed following a later spill of Nike Baby Sunray sandals.

These findings, augmented by additional years' observations from the Orbisons, would prove vital to understanding the subarctic Aleut Gyre and gyres in general. My intuition told me that the toys had completed a number of orbits around the Aleut Gyre. To test this hypothesis, Jim programmed OSCURS to trace the toys' trajectories through the twelve-year Orbison time line. His program had proven accurate for a single orbit. How would it do tracking four orbits?

Jim is an animator, and just as Walt Disney broke Mickey Mouse's movements down to twenty-four frames a second, OSCURS crunches daily weather data, calculates winds and currents, and shows where a sneaker, toy, or other piece of flotsam will arrive at ten-day intervals. The computer then plays the frames before your eyes, like turning the pages of a flipbook. Thirty-six frames represent one year. Over a period of months, I analyzed OSCURS's animations for seventy toys frame by frame—420 frames in total. (I could not track them further because a business Jim's wife had started—an assisted living facility—demanded all his energy. Since 2003 he has had little time to undertake any further animations, or even to go beachcombing.) I tested the OSCURS trajectories against the dates that toys beached at other locations in Washington and British Columbia. They held up well.

And OSCURS told us much more. It told us when the toys headed north into the Arctic through the Bering Strait, and when some would get diverted south into the subtropical Turtle Gyre. It also revealed three kinds of orbits: Some toys drifted once around the Aleut Gyre's periphery. Some made repeated orbits around the periphery—five in fifteen years. And others made loops—suborbits—within it. OSCURS showed that the Aleut Gyre (and by implication any gyre) is not a simple circle but a system of wheels within a great wheel, somewhat like a planetary gear.

For sixteen years, the toys kept stranding. Some drifted thirty-four thousand miles, far enough to circle the earth nearly one-and-a-half times. As the fifteenth anniversary of the spill approached, my calculations suggested that some would be approaching the North Atlantic after circling the Storkerson Gyre. Once again the toys generated worldwide attention. Media from Newfoundland to Virginia to Germany to Australia all called for stories on the wayward rubber duckies. The First Years, Inc. offered a $100 reward for the first confirmed Atlantic duck. NPR reporter Robin Young went to the beach near Boston and sought comments from beachcombers. Everyone she met was looking for ducks.

The ducks had become scientific as well as pop-cultural icons. On January 8, 2008, the *New York Times*'s Andrew Revkin reported on scientists desperately trying to puzzle out how much meltwater is running off beneath Greenland's ice sheet and where it's going. NASA engineer Alberto Behar told Revkin some unconventional methods were being considered to trace the water: "We had ideas to send rubber ducks down and see if they pop out in the ocean. They'd have a little note saying, 'Please call this number if you find me.'" Behar seemed unaware that he was proposing the equivalent of that hoary old research instrument, the message bottle. Toy ducks had supplanted bottles as the default drift marker.

All this hubbub over bath toys seemed ironic considering the lack of media attention for a much more unusual drifter—one of the most haunting in a tradition of "freedom bottles" from shuttered nations such as Cuba and Vietnam that harkened back to Yasuyori's stupas. In 1989, University of Washington oceanographer Richard Strickland celebrated his fortieth birthday by kayaking along Canada's Pacific coast. Answering nature's call,

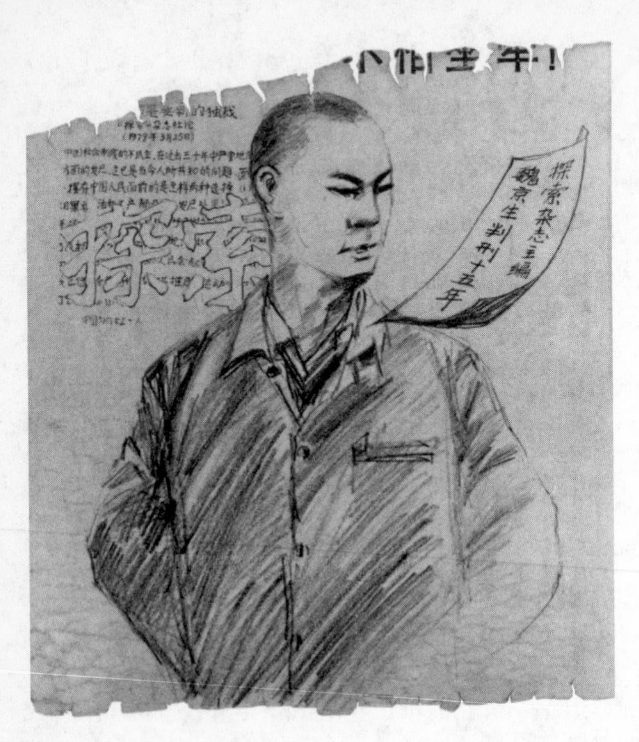

A Taiwanese protest flyer showing the imprisoned mainland dissident Wei Jingsheng, from a message bottle found on Vancouver Island.

he spotted a square, clear glass bottle containing multicolored leaflets on the beach. For three years it sat unopened on his mantle. One day, Richard and I met for lunch at a Seattle boat-workers' dive called the Fremont Dock. I asked if he'd ever found a message in a bottle. Yes, he said—and granted my father and me the privilege of opening it. Nothing—not tapping, fire, or scribing with a glass cutter—could break the seal. Finally the bottle broke.

Inside, a waterlogged sketch showed a young man standing beside Beijing's celebrated Democracy Wall, a poster calling for just that, and a scrap of paper falling from a judge's dais. "Don't be afraid of sitting in jail!" read the bold letters above.

The man was Wei Jingsheng, a twenty-nine-year-old electrician and dissident editor, China's most prominent political prisoner. In 1979, he was sentenced to fifteen years in the notorious *laogai* gulag for daring to urge that China add a "fifth modernization" to leader Deng Xiaoping's four economic reforms: democracy.

Where, when, and why was this salute to Wei launched? Every drifter holds clues to its origin. The Chinese characters molded into Richard's bottle translated as "Tobacco and Wine Monopoly Bureau of Nationalist Government on Taiwan." Inside, five of the leaflets were printed on both sides. Four were in full color and stamped with seven-digit codes, which suggested a large print run and sizeable bottle launch. Two clues suggested that the launch was undertaken in 1979, the year Wei was sentenced: One pamphlet mentioned the fifty-fifth anniversary of Taiwan's Whampoa Island Military Academy, founded in 1924, and a handbill noted that China's population now exceeded 900 million. The official estimate in 1979 was 906 million.

I concluded that Taiwanese activists had released thousands of bottles to alert mainland citizens to Wei's sentencing. They may even have used bottles to mock Deng: "Deng Xiaoping" can be translated as "little bottle." In his heyday, fans hung small bottles in trees as a gesture of support; when he cracked down on the democracy movement, smashed bottles littered Beijing.

The Aleut Gyre bore this particular bottle across the Pacific to Vancouver Island. It was discovered there in 1989, ten years after it was launched, around the time a far harsher crackdown unfolded at Tiananmen Square. This seemed to me an especially telling case of flotsam forensics—a confirmation of transpacific drift and a unique window into geopolitics and national psychology. I published an article on it, even longer than the one on the tub-toy migrations, in *EOS*. A representative of Amnesty International came to talk to me. But aside from Seattle's KING–5 News, which aired a vivid report, the media ignored the Wei bottle. They preferred flashy sneakers and cute tub toys to freedom drifters and fearless dissidents.

The sea is very tidy. When onshore winds blow, you can see the invisible hand of the floating world at work, sorting flotsam in both time and space. The wind pushes objects that rise above the water faster than those that are less exposed, and things wash up in sequences: Bic lighters one day, toothbrushes the next. Along the Washington coast, the first wash-ups to arrive are the airy purple jellyfish known as by-the-wind sailors (*Velella velella*).

Next come electric light bulbs, followed by the larger glass balls once used to suspend fishing nets. Riding lowest, and landing last, are the rolling-pin-shaped glass floats used to net octopus at the sea bottom.

Like birds of a feather, flotsam of similar wind resistance flocks together. One beachcomber, Vardon Tremain, came upon three beaches near the village of Tambor on Costa Rica's Pacific coast named after the flotsam they collected: Sandal Beach, Toy Beach, and Bottle Beach. The south shore of Maui has Lumberyard Beach, a hundred-yard stretch where driftwood piles up; very little reaches the adjacent shoreline.

Such inshore sorting marks the end of a process that begins thousands of miles up-current and echoes around the gyre. Pervasive though it is, it went unexamined until we tried to account for its effects with OSCURS.

Jim labored to estimate how the winds affected both the speed and route of the particular flotsam released in each container spill. He boiled windage down to two numbers. One is the wind factor—how much faster than surface water an object drifts under a given wind. The other is the deflection angle: Surface water has long been known to move at a forty-five-degree angle to the wind. But each type of flotsam moves at a different angle. So to get accurate readings of water movements, oceanographers seek to reduce the wind resistance of the drifters they release.

I got a vivid demonstration of this principle from a container spill that at first promised every bit as much good data on North Pacific currents as the Nikes and tub toys had delivered. That promise beckoned in a message from Alaska beachcomber Emily Becker: "On Kayak Island in the Gulf of Alaska, my friends and I were dumbfounded by the number of hockey gloves we found . . . We envisioned a boat load of hockey players lost at sea."

But no hockey players had gone missing. "We'll finger the culprit," joked Jim. And we did. On December 9, 1994, en route from Pusan, South Korea, to Seattle, the Cypriot-flagged containership *Hyundai Seattle* caught fire in the mid-Pacific. Powerless against storm waves, she drifted seven hundred miles while the Coast Guard evacuated her shaken crew. Salvagers towed her to Seattle, leaving forty-nine lost containers of Christmas merchandise behind. Eight months later, eight hundred miles west of Oregon, tuna-boat skipper Ron Anderson snagged several hockey gloves that had fallen off the *Hyundai.* When Anderson docked at Newport, a NOAA fisheries inspector spotted four gloves on his bow and brought them, still dripping, to Jim Ingraham. Flotsam retrievals that far out to sea were rare, so Jim cranked up OSCURS to parse out this one. It showed that the winds had blown the gloves 20 percent faster than the surface currents—revealing the windage factor Jim needed to forecast their landfall.

Month by month, the NOAA staffer Lynn De Witt emailed navy grids of North Pacific atmospheric pressures to Jim so he could update his forecast of the gloves' progress. We waited patiently until January 14, 2006, more than a year after the spill, when OSCURS forecast that the gloves would land at Barkley Sound, on southwest Vancouver Island. "That's today!" I exclaimed. Just then, the phone rang and a beachcomber named Alex Welcel told us he'd found fifty or sixty hockey gloves along a three-mile stretch of beach on Barkley Sound. OSCURS had passed another blind test.

As the year progressed, thousands more gloves beached from the California-Oregon border to the Queen Charlotte Islands. I set out to trace them and, after dozens of phone calls, struck pay dirt when the Hong Kong–based exporter who'd ordered them sent me his original invoice. It disclosed that 34,300 hockey gloves had tumbled off the burnt-out *Hyundai,* along with 34,000 Avia sneakers.

The two behaved in intriguingly different ways at sea. The sneakers washed up a full two months after the gloves had completed the two-thousand-mile crossing. They were in much better shape than the gloves, whose fingers were badly frayed. At first I thought the sneakers were just better crafted. Then I compared how the two floated. The shoes rode soles up, with their fabric bodies protected underwater. The gloves, by contrast, rode like mini-icebergs, their fingers sticking up to be bleached by the sun and pecked by seabirds.

That also explained why the gloves traveled so much faster than the shoes: The wind slid over the sneakers' smooth, low-riding soles but turned the gloves' upright fingers into sails. If the gloves traveled all the way around the Aleut Gyre, the wind factor could shorten their orbital period by as much as eight months, down from 3 to 2.3 years. This is a serious statistical bias; I had to discard the gloves from my calculations of the gyre's orbital period.

Over thousands of miles, even subtle differences in wind resistance and water drag can sort out drifters. I wondered if the toys collected at Sitka might show some sorting. Sure enough, the hamperful Dean and Tyler Orbison beachcombed near Sitka held thirty-five ducks to every twenty-six beavers, twenty-one frogs, and eighteen turtles. I could only guess where the missing frogs and turtles had gone, but I saw now that wind-blown sorting affected even similarly sized objects made of identical material—if their shapes differed. Turtles and ducks were apples and oranges as far as the ocean was concerned.

Even the slight difference between left- and right-handed gloves could send them on different sea routes. The sea has eons of experience sorting left from right. Geologist-beachcomber Peter Hilary Alexander Martin-Kaye, who was as obsessed with shells as I am with shoes, counted 895 specimens of a clam called the Royal Comb Venus along Trinidad's Cocos Bay. A curious row of spines runs up one side of each Royal Comb, accentuating the asymmetry between its "right" and "left" shells. The sea sorts them accordingly. At one end of the bay 87 percent of the shells washed up were rights; at the other end, thirty miles away, only 11 percent were.

Left- and right-handedness extends to many other sea creatures, from flounders to the sailing jellyfish *Velella velella*. It is a highly efficient evolutionary adaptation that helps species disperse more widely. For example, female sea skates have two tubes—the equivalent of two uteruses—through which they deliver their young, enclosed in pouches. The two tubes extrude differently shaped "left" and "right" pouches. These scatter like shoes and clamshells, upping the odds that some of each mother skate's progeny will survive.

"Each beach has its own peculiarities," Nick Darke once told me. "Porthcothan Beach behind my back gate on the North Cornwall coast accepts three times more left-handed gloves than it does the right." Steve McLeod found seventeen left-handed gloves and just six rights on his Oregon turf.

As with gloves, so with shoes. Walking the beach near Cocoa Beach, Florida, one day, I found a dozen right-footed shoes to one left. Mardik Leopold, a researcher at the Netherlands Institute for Sea Research, examined hundreds of shoes washed up around the North Sea. He found that rights favored the Shetland Islands on the sea's west side and lefts collected along the eastern shore. He also conducted wind-tunnel tests of sneaker hydrodynamics and found that water flows around shoes just as air flows around an airfoil, tending to nudge them in one direction or another. And the difference between left and right feet is enough to change that direction.

A Texas nurse and beachcomber named Tally Calvert patiently tallied a year's worth of beached footwear on the west end of Galveston Island, Texas, which she then hung on a backyard fence she dubbed the "Wall of Lost Soles." She found that lefts outnumbered rights 120 to 51 and predominated in every category, from boots to high heels. The odds against such a lineup happening randomly are greater than ninety-nine to one.

So where are the missing rights? Doubtless on some other beach where the currents favor them. Or the missing lefts? That question becomes more fraught when not just shoes but entire feet start washing up.

5. Coffins, Castaways, and Cadavers

You sea! I resign myself to you also. . . . I guess what you mean,
I behold from the beach your crooked inviting fingers. . . .

—Walt Whitman, "Song of Myself"

Buoyed up by that coffin, for almost one whole day and night,
I floated on a soft and dirge-like main.

—Herman Melville, *Moby-Dick*

Another flotsam-fed media frenzy erupted in spring and summer 2008 over the mystery of the floating feet in British Columbia. Between August 2007 and June 2008 five feet—four right and one left matching one of the rights—were found along the Georgia Strait, which separates Vancouver Island from the Canadian mainland. All were wearing sneakers; four were size 12s. A sixth foot appeared, but proved a gruesome prank: what the authorities described as "an animal paw" in a sneaker. DNA samples from the others matched one missing person but not any other known missing persons. Deliberate dismemberment seemed unlikely as police did not find cut marks on the bones.

Once again the world's media descended on the Northwest coast, enthralled by a floating mystery. Again they came to me seeking answers; distraught individuals even called seeking their missing loved ones. But I could offer only speculation and general insights. Not that the subject was an unfamiliar one; over the previous two decades, while I traced the sneakers' and bath toys' movements around the Pacific, I'd embarked on a parallel, perhaps equally bizarre-sounding research track. Call it forensic flotsamology: the study of the movements and behavior of human bodies and body parts at sea. Cliff Barnes always taught us never to ignore good data, no matter where they came from. And the lode of forensic data was wide and, in some cases, uniquely revealing.

There's nothing unusual about body parts washing up. Yanked by currents, battered by obstructions, and nibbled by scavengers, bodies commonly disarticulate as they drift. Heads usually come loose first, prompting the irrepressible Jim Ingraham to observe, "That could save somebody's neck in court." Feet, legs, torsos, and arms follow—ten or so separable parts per body. This run of mostly right feet, in the absence of other parts, seemed to defy all odds. It would be normal for the lefts to wash up elsewhere—but where? And what about the other parts?

For now, the sneakers on the feet seem key. I do not know if these had "air" soles, but the foam in ordinary shoes would help feet float to the surface. And because sneakers float sole up, they would protect their contents against scavenging birds.

Seattle mystery writer Kathrine (K. K.) Beck shared some insight with us on that score: "I once knew a troubled young man who fell into an icy fjord in Alaska while stealing a boat after a long shift at a fish cannery and a case of beer—a really tragic thing. His feet bobbed up much later and many miles apart—one of them snagged by a fly fisherman from a riverbank. Apparently, while the sea used to swallow the dead, now, because of buoyant athletic shoes, feet wearing them routinely detach themselves after a certain amount of crab activity on the bottom. So apparently it isn't that amazing that these things wash up, and the authorities in Alaska told his parents that it happens all the time."

It was a tip from my mother that got me started on forensics, just as she'd started me on flotsamology. In February 1988, two years before the sneaker spill, she showed me a newspaper article about a body that had washed up just below Alki Point, where Seattle juts out farthest into Puget Sound. I went on to piece together the rest of the story. Two nights earlier, three soldiers from nearby Fort Lewis had walked out onto the spectacular

Tacoma Narrows Bridge, the successor span to the notorious "Galloping Gertie" that twisted apart in high winds in 1940, providing generations of high school students who've watched the film of its collapse an unforgettable demonstration of oscillating wave motion. The bridge's second edition had no such violent reactions to winds, but its twenty-story height did attract jumpers.

The three soldiers stood ready to jump. Two said later that they were only kidding, but their comrade leapt. Four seconds and 224 feet later, he hit the cold water at eighty miles an hour, the same speed reached by jumpers from the Golden Gate Bridge. The impact ripped his jeans from his body and ruptured his internal organs.

The newspaper story noted that King County medical examiner Bill Haglund had examined the body. I called Haglund and offered to simulate the drift of the jumper's body in the UW Oceanography Department's hydraulic model of Puget Sound (Cliff Barnes's brainchild, a magnificent replica that we practical ocean types rallied around every time the university tried to trash it; after half a century it still beats computer models, especially at simulating tidal eddies). At the time, Haglund was hot in pursuit of the Green River Killer; when I visited him in his office, I saw a skull and other grisly evidence from the killings. But he liked my idea; perhaps he was glad to get a break from that ghastly case. We dropped a plastic bead representing the jumper in the model and watched it progress about one second per actual hour of drift. It showed that after the jump, the body drifted south for an hour or two, then headed north through the Narrows and Colvos Passage, rounded Blake Island, and turned east. Before beaching, it circled in a flood tidal eddy in the south lee of Alki Point.

We wrote up this case for the *Journal of Forensic Science*, and it proved a popular paper; this was the first time we knew of that a hydraulic model had been used to trace a human body's drift. As time went on, Bill brought me in on other cases and I traced the drifts of many human remains. In 1995, I was called to provide expert testimony on the drift of a severed head in a sensational Oregon murder case, in which a woman named Linda

Stangel was convicted of pushing her boyfriend off a three-hundred-foot seaside cliff near Cannon Beach, Oregon. Steve McLeod took me there to examine the site.

Sometimes it seemed I could not escape the gruesome subject. My office window faced the Aurora Bridge, Seattle's answer to the Golden Gate, from which more than two hundred people have leapt into the Lake Washington Ship Canal. Just as I was revising my account of the Tacoma jumper, I looked up to see a man plummet past my window. Several weeks later, another man landed in our parking lot.

It could have been worse. I could have had a view of the Golden Gate Bridge. By 2005, when the authorities stopped releasing counts, more than twelve hundred unhappy souls had jumped from it—not counting the two drifters who provided my UW classmate John Conomos, another Cliff Barnes protégé, some extra data when he charted San Francisco Bay's currents in 1971. They were tong gang members whom rival gangsters had killed at the Presidio and dumped into the bay. At first they sank into the deep southward-flowing currents. Then, off Oakland International Airport, the bodies bloated, as bodies will, and rose to the surface. They continued drifting, more than a mile per day, and eventually beached about thirty-five miles below the bridge.

John had obtained one of the oldest forms of oceanographic data. Fourteen hundred years ago, the propensity of bodies to bloat and float revealed the currents of the Bosporus, the historic strait that divides Europe from Asia. In 617, the Byzantine emperor Heraclius repelled Persian invaders who had fought all the way to the Bosporus, opposite his capital Constantinople. He had the Persians' corpses beheaded and thrown into the strait. The bloated torsos floated south on the strong surface current, while the heads sank and rolled north on a deep, even stronger undercurrent. Eventually their teeth drifted into the Black Sea. Boatmen realized that if they hung stone-filled baskets in that undercurrent, it would drag them north against the surface current. Thus was revealed the dual flow—surface water in one direction and deep water in the opposite—that oceanographers call an estuarine current pattern.

Just as I became interested in human drifters in the 1980s and early 1990s, I was also trying to figure out what would become of oil spilled on Northwest waters. One scenario haunted me, though my colleagues thought it unlikely. Large tankers carried oil from Alaska to the refineries at Cherry Point, about one hundred miles north of Seattle. I worried that tankers entering and exiting the Strait of Juan de Fuca might collide. Violent winter storms were frequent there, and they could stir up waves matching or exceeding any in the worst Gulf hurricanes; I had located three records of hundred-foot waves just to the north, off the coast of British Columbia. Such a storm might drive floating oil more than one hundred miles into Puget Sound and the other inland waters. But no one had ever released drifters under such circumstances, so federal and state regulators dismissed this scenario as speculative.

Then, in a brief account by the historian James A. Gibbs, I came across a clue that led me to pertinent drift data. The source was a tragic shipwreck more than a century earlier. On November 4, 1875, the paddle-wheel steamer SS *Pacific* left Victoria for San Francisco with two to three hundred passengers, many of them miners heading off to spend what they'd gleaned from the Cassiar gold fields. The clipper ship *Orpheus* meanwhile sailed north along the Washington coast. The two vessels collided in the dark off Cape Flattery, at the mouth of the Strait of Juan de Fuca, an infamous watery graveyard. The *Orpheus* effected hasty repairs and sailed on. Out of sight, the rundown *Pacific* sank in thirty minutes. All but two passengers drowned.

Over the next month, three powerful storms scattered the *Pacific*'s dead along a hundred-mile swath from twenty miles off the ocean coast to eighty miles up the strait at Victoria. Searchers found debris and bodies stretched along a sixty-mile line of wrack where fresher inland waters met the salty Pacific. I realized that if corpses could travel that far on the currents, surely oil could too; and if it got that far, submerged oil would drift into Puget Sound.

I found some confirmation when I studied the December 1988 *Nestucca* oil spill off Grays Harbor in Southwest Washington. Some of the

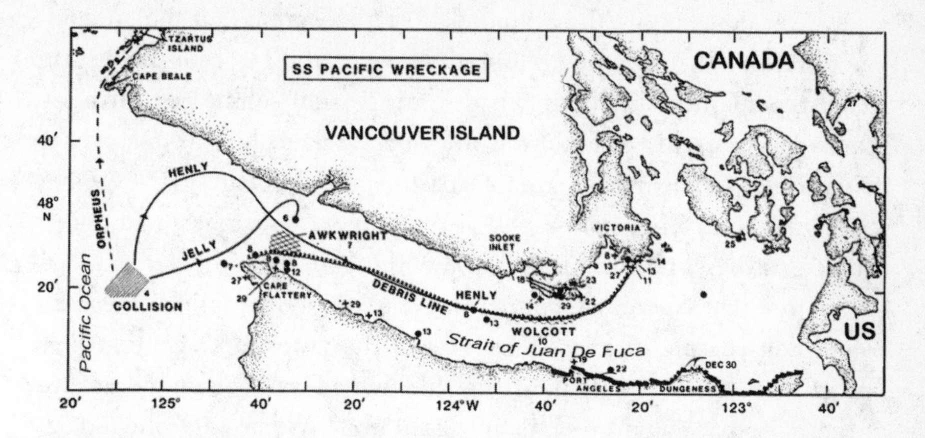

Drift of wreckage from the SS Pacific *from Cape Flattery into the Strait of Juan de Fuca, indicating a likely path for spilled oil. Dots on shoreline show where and on which dates debris or bodies washed up in December 1875.*

231,000 gallons that leaked from the barge *Nestucca* drifted north, turned the corner at Cape Flattery, and washed up along the same stretch of the strait as the *Pacific*'s victims. But even that was not enough of a wake-up call for the authorities, who continued to pretend that oil would not travel up the strait.

For about ten years, from the late 1970s to the late 1980s, I had done oceanographic work for two proposed oil pipelines, Northern Tier and the Trans Mountain, both of which would have run eastward from the Port Angeles area, under the Strait of Juan de Fuca to the oil ports at Anacortes and Cherry Point and points beyond. Either would have obviated the need for Alaska tankers to twist through the narrow channels of Washington's inland waterways and greatly lessened the danger of a massive oil spill in the strait. But neither pipeline was ever built; environmentalists raised many concerns and the refineries balked, preferring to receive their oil the way they always had, via tankers. And so I continued lecturing on the lessons of the *Pacific*—to deaf ears.

I became increasingly worried about the danger of submerged oil infiltrating Puget Sound—fearful that the strong tidal currents might suck the oil downward in Admiralty Inlet and into the deep estuarine water layer flowing into the Sound. Colleagues working for the oil companies repeatedly urged me to drop the subject. But I was not the first to sound this alarm; Cliff Barnes had warned about submerged oil more than a decade earlier. While I was studying the issue, he gave me a letter he'd written to the Washington Department of Ecology in November 1974. It recounted how University of Washington researchers had traced submerged diesel oil from a spill at the Texaco refinery dock near Anacortes through Guemes Channel and turbulent Deception Pass into Skagit Bay—despite the fact that most of the water in Deception Pass flushes in the other direction. Meanwhile, "no oil was seen on the surface of the bay" that could give this intrusion away. Worse yet, Cliff noted, Puget Sound's main basin would be "much more vulnerable" to such a threat than those waterways because of its slow flushing and large intake of deep water.

I entered this alarming letter, from an emeritus professor who knew more than anyone about the Sound, into the report I prepared for the U.S. Environmental Protection Agency. It disappeared into the bureaucratic maze, like the lost ark at the end of the Indiana Jones movie. As far as I know, Cliff likewise never received a reply. My work on the *Nestucca* spill and the SS *Pacific* received the same response: none whatsoever, save from treasure hunters seeking the gold the *Pacific* supposedly carried. Despite endless political lip service, no one has been so avid to protect the real gold of our fragile waters.

I took up forensic beachcombing in 1988, four years before the great toy spill. Twenty years later I had compiled nine notebooks on drifting human remains and nine on the related (as we'll later see) subjects of floating volcanic pumice and the origin of life. All at once, in the mid-1990s, questions of life and death took on a much more personal significance. The three men who'd done the most to shape my life neared the end of theirs. Cliff Barnes finally succumbed to Alzheimer's disease in 1995. Colon cancer overcame Akira the next year. And my father was in the last round of his long battle with Parkinson's disease and many other ailments.

But Dad remained passionately interested in my work and supportive of my mission. When Parkinson's made his hand shake too much to continue drawing the illustrations for my articles, he turned to writing poetry about the sea and its drifters. When glaucoma fogged his eyes, he dictated his poems to my mother.

> *Oh, continental drifter with rings of history*
> *Set adrift by disaster and mishap . . .*

he wrote in his last poem.

> *You have provided mankind*
> *with perfumes, incense, staffs, and canes*
> . . . .
> *You have served us well.*

From 1980 on I stopped by my parents' house almost every day for an hour or so to sip wine and talk. One day in 1996, Dad broached an idea that would mark another turning point—that I publish a newsletter about beachcombing, flotsam, and currents. Jim White, the physical therapist my mother hired once or twice a week to ease my father's pains, happened to be there for lunch. He offered to lay out the newsletter on his computer. Mom said she could manage the mailing list. All I had to do was write it. And so the *Beachcombers' Alert!* came to be. We mailed the first issue in July 1996. Dad never got to see his brainchild; by then glaucoma had stolen his eyesight, and he passes away later that month.

The *Alert* soon took on a life of its own. It became the node of a worldwide network of informants—an army of beachcombers scouring the shore from Iran (one of the shah's former bodyguards receives it) to Pitcairn Island, seeking flotsam evidence of the gyres' movements.

Beachcombers' Alert!™

Since 1996

Vol. 9 No. 2 • October - December 2004 • Curtis C. Ebbesmeyer, Ph.D.

Beachcombing Science

Dean Orbison with turtles, ducks, beavers, and frogs. *After 12 years adrift, the frogs and turtles remained true-blue to their original colors (green and blue, respectively), whereas the ducks and beavers faded from yellow and red to white, respectively.* ***Dorothy Orbison*** *photo*

Beachcombing Science From Bathtub Toys by Curtis C. Ebbesmeyer

Twelve years and counting — the saga of the tots' tub toys continues. On January 10, 1992, 28,800 turtles, ducks, beavers and frogs packed in a cargo container — called *Floatees* by the manufacturer — splashed into the mid-Pacific, where the 45th parallel intersects the International Date Line (44.7°N, 178.1°E). During August-September, 1992, after 2,200 miles adrift, hundreds

Continued on page 2

The Call of Our Dust

46,000 *by* ***Richard Lang*** *and* ***Judith Selby.*** *After years adrift, the bathtub toys (left) disintegrate to bits of floating plastic. "The California Coastal Commission reports 46,000 pieces of visible plastic floating in every square mile of the ocean. This shocking fact along with our inability to visualize the magnitude of that quantity compelled us to count and attempt to exhibit 46,000 plastic pieces collected from Northern California beaches," wrote the artists. "When we committed to this work, we had no idea of the enormity of the undertaking. To date, we continue our task. On a hundred wires, we strung 4,600 pieces of plastic to simulate the colorful bits of plastic floating in 1/10 of a square mile of sea."*

Multiplying 46,000 plastic pieces per square mile by the total area of the ocean yields 6.4 trillion pieces of visible plastic afloat on earth, equivalent to a thousand pieces per earthling. The U.S. Environmental Protection Agency estimates that Americans annually make 910 million trips to the beach. If the ocean transported all these bits to U.S. shores, Americans could rid the sea of visible plastic by picking up seven thousand bits per trip. Copyright photo by Richard Lang and Judith Selby.

Continued on page 4

Beachcombers' Alert! *front page, showing master beachcomber Dean Orbison with washed-up toys and plastic dust on the sea.*

Before he passed away, my father opened another window. He shared a story he'd learned in his youth, perhaps the earliest account of the floating world, which set the stage for many tales to follow.

Dad had set out to write his own book—science fiction, embracing the entire universe—but Parkinson's disease stopped him from finishing it. Nevertheless, through all the years of tremors and hallucinations, he retained sharp images of his youth in Chicago in the Roaring Twenties, when his widowed mother made bathtub gin and lived in the storage rooms above Pop Hirstler's speakeasy, past the railroad tracks. The times and the neighborhood were rough; a stray bullet from the street once passed within a foot of Dad's crib. Delivery runs of hooch took off regularly from Pop Hirstler's, and Dad went along on one while he was still a boy. Most of the time he played in the alleys. Even then he was an avid reader, and he never forgot the first book he ever owned, salvaged from a trash bin. It was Albert Pike's *Morals and Dogma*, the bible of the Freemasons. Pike's book led my father to become a Mason himself, and after fifty years as a lodge member he still read from it, like a gyre returning to the shore from which it started.

Dad returned again and again to a story Pike took from one of the foundation myths of ancient Egypt. It told of the posthumous voyage of a legendary pharaoh (perhaps a real predynastic king) named Osiris. Osiris's malicious brother Seth envied his fame, his power, and, especially, his beautiful wife—their sister Isis—and hatched a clever plot. At a royal banquet, Seth offered a gold-covered sarcophagus to the man who fit best in it. When Osiris lay down in the casket (carved to his dimensions, of course), Seth's minions nailed it shut, sealed it with molten lead, and heaved it into the Nile.

The current bore Osiris's casket north from the Nile's mouth past Sinai, Gaza, and Israel to the Phoenician city of Byblos, and wedged it in the crook of a tamarisk tree. Isis followed the currents, recovered Osiris's body, and revived him just long enough to conceive a son, the falcon-headed god Horus, who would take vengeance on Seth. But when she brought the corpse back to Egypt for burial, Seth cut it into fourteen pieces and scattered them across the land—a gesture tied to the planting of the Egyptians' crops, and hence to the annual Nile floods that irrigated those crops. Faithful Isis retrieved and reassembled all the pieces save Osiris's penis, for which she fashioned a replacement. One account says it was then that she conceived Horus.

Osiris's story probably has its roots in the very real movements of water, men, and goods along the eastern Mediterranean. Archeological excavations show that Pharaoh Sneferu obtained timber—Lebanon's famous cedar—from Byblos some 4,650 years ago. Mariners returning to Egypt could have told Isis that her husband would wash up at Byblos. The coffin's drift concurs with contemporary data, including a number of satellite-tracked drifters, which show that the Nile's effluent turns eastward and northward, following the shore. After a great flood in 1938, floating islands and drowned cattle washed up as far north as Byblos. The ancient Egyptians surely observed the same processes.

All great stories repeat themselves, Karl Marx might say if he were a flotsamologist—the first time as myth, the second time as accident report and ghost story. Osiris's voyage was reenacted by another pharaoh's remains just 188 years ago. According to the irrepressible Dutch artist and beachcomber Henk Noorlander, in 1821 a Prussian diplomat looted a tomb now believed to be that of the pharaoh Djoser, who died around 2300 BCE. The ship bearing the booty to Germany foundered in Elbe Bay, and Djoser's granite sarcophagus was buried again. Ever since, when western storms battered Elbe Bay, the old folks would whisper, "It's the pharaoh's curse."

Osiris's story, together with the drifts recounted in the Icelandic sagas and *Tale of the Heike*, led me to suspect that every legend carries grains of truth. Floating tales bracket human life in a fateful symmetry. Babies float into the world, and the dead return to their origins in the floating world. For eons, young mothers have anguished over what to do with infants born out of wedlock or in other inconvenient circumstances. Again and again, history and legend deliver the same answer: Send them floating off and entrust them to the gods. The most famous and fateful of these waterborne babes is the son of Jochebed, abandoned in a pitch-covered basket in the Nile reeds. He becomes Moses, meaning "drawn out [of the water]."

Moses's abandonment echoes often in the lore of the Middle East and elsewhere. "Put your child in the ark and let him be carried away by the river," the Qur'an advises. A surprising number of those thus carried away survive the trial by water to become extraordinary figures. Romulus and Remus drift in a basket down the Tiber River. Two blind Burmese princes are likewise set adrift. Thebes's wrathful King Acrisios casts his daughter and her infant son, the future hero Perseus, out to sea in a casket. In *Beowulf*, Scyld of the Sheaf drifts across the waves to the Danes. In the Indian epic the *Mahābhārata*, the warrior hero Karna, the immaculately conceived son of the sun god, is nevertheless cast downstream by his mother.

The great Mesopotamian king Sargon, born of a similarly shamed mother, recounted that "she laid me in a vessel of rushes, stopped the door thereof with pitch, and cast me adrift on the river . . . The river floated me to Akki the water drawer, who, in drawing water, drew me forth."

In seventh-century China, a pregnant wife abducted by a bandit gave birth in captivity but sent her child floating to freedom down the Yangtze, after recording his origin in her own blood. A monk found the bucket and raised the boy as his disciple, and he in turn grew up to be known as T'ang-seng, the Floating Monk.

Taliesin, the earliest-known Welsh poet, was cast into the sea in a skin bag. Several of the holy men of Wales and Ireland also began their lives as flotsam: Saint Cenydd, set adrift in a cradle; Saint Kentigern, whose pregnant mother was cast out to sea in an oarless boat. Fifteen hundred years ago, goes a tale still told in Ireland, a cow was seen licking a box washed up at Kilcummin Roads. Finally a woman became curious and opened it. The baby inside grew up to become St. Cummin, a great worker of miracles.

The Buryat people of Mongolia have passed down an account of a distinctly unholy luminary who arrived by similar means. Once, it seems, a khan had two wives. The elder tricked her husband into ordering the younger to cast her newborn child into the lake in a pitch-covered cradle. Just as it stranded and the infant kicked open his little ark, a lark sang "Shin, shin!" And so he was named Shingis or, as we spell it, Genghis Khan.

Sometimes the gods of the floating world still smile on young innocents. In 1908, eight thugs kidnapped eighteen-month-old Renée Nivernas and held her on a yacht moored off Marseille. Before the ransom could be negotiated, a storm sank the yacht and drowned the abductors. Little Renée, asleep in a cradle made of packing crates, washed safely ashore twenty miles away.

It is wise nonetheless to be wary of currents bearing gifts—as Maarten Boon's grandfather and his tippling cronies discovered when they scraped the bottom of their barrel. Boon, a blacksmith on the Dutch Island of Texel, is heir to a long wrecking tradition; he once ate porridge for three months from cans his father had beachcombed. That was nothing next to what his grandfather found. "One day he found a two-hundred-liter barrel on the beach," recounted Henk Noorlander, who passed Boon's story on to me.

"He rolled it into the dunes, pulled out the stopper and sipped pure alcohol! And so he ended up with four hundred liters of hundred-proof spirits. Then he invited everybody for a drink."

Thereafter, grandpa trekked into the dunes whenever his jug ran dry, till not a drop came out of the barrel. He grabbed an ax and cracked it open—"and found a monkey preserved in pure alcohol," presumably a zoological specimen lost off a ship. Wherever it came from, for Grandfather Boon there was nothing more fun than a barrel of monkey.

Pickled primates aside, deceased drifters inspire sorrow more often than amusement. Usually the wind, waves, and fishes provide a rough burial at sea, but occasionally skeletons and mummies make it to shore after epic voyages. The seventeen-foot Boston whaler *Sarah Joe* began drifting when her motor conked out during a violent storm in February 1979 off Maui's east coast. Ninety days later she washed up at Taongi Atoll in the Marshall Islands, two thousand miles to the west. No trace remained of four of the five fishermen who'd been aboard. But the bones of the fifth reached the atoll, and sympathetic fishermen gave them an onshore burial.

In 1975 the twenty-foot sloop *Grace A. Ghislaine* was retrieved off Mexico's Yucatán Peninsula with two skeletons sloshing in the bilge. With sails unfurled, she'd drifted two thousand miles from Martinique in two months, half again as fast as the *Sarah Joe*.

Ghost ships and coffin boats, their crews dead or missing, have surely drifted the seas since sailors first took to them. They have inspired great and not-so-great art since the dawn of literature: Homer's *Odyssey*, Coleridge's "Ancient Mariner," Poe's "MS. Found in a Bottle," Stoker's *Dracula*, Hollywood's *Pirates of the Caribbean* movies. The captain in Wagner's *Flying Dutchman* comes ashore seeking love every seven years—just about the orbital period of the Pacific Ocean's vast Turtle and Heyerdahl gyres.

Usually such death drifts are isolated events. But every decade or so a fleet of makeshift craft shoves off from one unhappy land or another, crowded with hungry migrants or desperate refugees—many of whom never make it across the waters. In the 1970s and 1980s it was the million Vietnamese boat people. Most recently it's been West African fishermen displaced by European factory fishing and their neighbors, struggling to get to the Canary Islands and Europe. In 2005 one of their boats broke down and drifted across the Atlantic to the Barbados still bearing a dozen mummified bodies.

In between, in the early 1990s, the "freedom flotilla" poured off Cuba on anything that would float. On July 31, 1994, two young men, Omar Granda-Rosales and Hilberto Samuel, shoved off the Isla de Piños off Cuba's south side. They'd constructed a seven-foot boat with aluminum from the Granda-Rosales family's roof and sails sewn out of flour sacks from the bakery where Granda-Rosales worked. It was Granda-Rosales's fourth escape attempt; his mother, who'd made his pants from burlap, said he'd just been released after a year in prison for plotting his third.

Granda-Rosales and Samuel had the bad luck to shove off at just the wrong point in the life cycle of the Loop Current, one of the few currents whose name actually describes it. The Loop Current pushes north from the Yucatán every year or so, tracing a bulge like a pregnant belly. Early in the pregnancy it flows north through the Yucatán Channel, rounds Cuba, exits the Gulf of Mexico stage right between Cuba and Florida, and seeds the Gulf Stream. As gestation proceeds, the loop bulges farther north, almost to the Mississippi Delta. Then, in an instant, the water breaks and the bulge snaps, splitting off into a two-hundred-mile-wide eddy. The Loop Current meanwhile resumes its normal course around Cuba.

Granda-Rosales and Samuel counted on drifting around Cape San Antonio at Cuba's western tip and catching the speedy loop that had carried so many of their compadres east to Miami. But the current had other ideas.

The Loop Current's whims are of surpassing interest to the oil industry; they can affect conditions in the platform-studded Gulf dramatically, even catastrophically. The current carries a deep, immense load of warm tropical water into the Gulf. If the right storm passes over it or one of its eddies, it can draw enough energy to grow into a shattering hurricane. Thus it was that the eddy that oil industry oceanographers had dubbed Fast Eddy fueled Hurricane Juan in 2003.

The industry had meanwhile moved into ever-deeper waters. When I started at Mobil in 1969, one hundred feet was the normal water depth for offshore platforms. By the late 1980s the industry was planting platforms more than two thousand feet deep. Bob Hamilton continued to be the industry's go-to guy for offshore measurements, and around 1977 I published a paper on currents he'd measured entitled "Strong, Deep Currents in the Gulf of Mexico." We found that when those currents surpassed two knots, as they typically do in a loop eddy, they could make the oil platforms vibrate till they thrummed like flagpoles in a stiff breeze. Nevertheless, the paper didn't create much of a stir, and the oil companies continued blithely operating in ever-deeper waters.

In February 1989, the industry got the message with a bang. The Loop Current spun off mighty Nelson Eddy, whose four-knot current blew a rig off its drill hole and sent it floating away, trailing pipes and cables like a giant jellyfish. The cost was two months' delay and $20 million in repairs. Many offshore rigs measured the currents Nelson slammed into them. Afterward the oil industry decided to assemble all this data and reconstruct the eddy's catastrophic progress—just as when we'd reconstructed the waves generated by Hurricane Camille and persuaded the industry to beef up its oil-rig standards. But now we were dealing with currents from a very different oceanographic process.

I was fortunate, or unfortunate, enough to submit the winning bid to conduct this review. It proved to be a nightmare. I had to assemble reams of current data of varying quality, taken at different depths over different periods with different instruments for many different oil companies. We developed a model that was much more complicated than the one then in use for designing oil platforms. Industry oceanographers wanted to simplify it. They pushed back, but we pushed ahead.

At first we thought that we could understand the eddy by looking at data taken at five-day intervals. Then I received a fax from Scott Glenn, the gifted computer modeler on my Nelson Eddy team. It showed that Nelson actually changed substantially even when viewed at two-day intervals. I groaned; I could see that we would have to run our elaborate model at one-day intervals. I faxed Scott back and he good-naturedly ran the model again.

Now we could see much more clearly how the giant eddy proceeded. It moved by fits and starts, stalling for a week and then changing position and shape. After it broke off from the Loop Current it seemed to bounce off the continental shelf, its ellipsoid shape springing like a rubber ball thrown against a wall. We kept the model running till Nelson moved off to the west and no longer endangered any rigs. Several years later, it died in the eddy graveyard of the Western Gulf of Mexico—off Matagorda Island, where so much flotsam goes to strand.

Nelson Eddy encouraged the industry to include current surveys in its drilling plans. To avoid a reprise, the oil companies now jointly fund two of my old colleagues, Patrice Coholan and Jim Feeney, to track the progress of Loop Current eddies with drifting buoys. They summarize their findings in their weekly *Eddy Watch* newsletter.

Omar Granda-Rosales and Hilberto Samuel did not have the benefit of this information when they shoved off from the Isla de Piños on July 31. Two days ealier, *Eddy Watch* had reported that the Loop Current was fully pregnant and ready to calve an eddy. As Granda-Rosales and Samuel rounded Cape San Antonio, satellite infrared images showed strong currents nearby, and Patrice discovered that the Loop Current had just shed Eddy Yucatán. Granda-Rosales and Samuel were condemned to swirl around this infant eddy while their water and food ran out. After forty-five days, they'd drifted within sixty miles of Louisiana. Crewmen spotted their boat near an oil-production platform. The ten-foot waves had rocked Samuel's body overboard. The next day, the Coast Guard recovered the seven-foot vessel and Granda-Rosales's mummified remains.

A thousand years ago, the Vikings followed a similarly sized eddy to new homes in Iceland. The crews of Japanese junks rode the Turtle Gyre—a much bigger eddy—around the North Pacific; some may have survived and found new homes among the American tribes, but most surely met a grimmer fate. When Omar Granda-Rosales and Hilberto Samuel trusted the current, they found only death.

At least their fate is known. In the early 1990s, the remains of eight to ten rafts and several small boats washed up on Cumberland Island, Georgia. Money, newspapers, and other effects from Haiti were found aboard them. When contacted, the Coast Guard asked if the boats or rafts bore any orange markings, which would have been affixed to indicate that their passengers had been rescued at sea. None did.

All along Florida's eastern coast, Cuban freedom drifter rafts washed up. One was a patched-up truck-tire inner tube into which a desperate refugee had strapped himself. He survived, but by the time the Coast Guard rescued him, the fish had begun eating him alive. A Melbourne Beach artist named Wayne Coombs thrust three oars through the tube and set it out as a silent memorial. Anything the sea touches becomes art, he told me.

One raft of bamboo timber logs beached near Cape Canaveral, within sight of a space shuttle perched on the launch pad. The human race may conquer outer space, but we're still bound to the floating world.

One of matter's most fundamental traits, specific gravity, is commonly described in terms of floating. A substance with a specific gravity of less than 1 floats on fresh water. One whose gravity is greater sinks. Human beings and bowling balls straddle that value, and with both the question comes up more often than you might think. A surprising number of bowling balls wash up on beaches. I wondered why until I learned at a beachcombers' fair that they are amateur cannon makers' favorite ammunition. Some have also been launched from aircraft carrier catapults. Those weighing twelve pounds or less (a midrange ball) float, while heavier models sink. I still wonder how J. D. Salinger came to title one of his stories "The Ocean Full of Bowling Balls." Perhaps he, like Poe, possessed an uncanny oceanographic sense.

As with bowling balls, so with human heads and bodies. Charting the oceanic progress of both in death inquests, I've noticed a striking pattern:

About half of them float and half of them sink. (In the only scientific study on the subject, 69 percent floated and 31 percent sank.) It all depends on their ratios of muscle, bone, fat, and fluid. Bob Hamilton once saw two University of Washington football players take a swimming lesson ordered by their coach. One, an offensive lineman, was a natural floater; he could barely propel himself underwater. The other, a running back and natural sinker, splashed about but could not stay afloat.

We resemble the fluid from which life emerged more than we know; our specific gravities match that of water within a few percentage points. That fact led to the peculiar but once-pervasive practice of trial by water. Until the nineteenth century, political and religious authorities looked to the sea or the nearest river to sort the guilty from the innocent. It was a perilous ordeal. If the water rejected accused criminals or suspected witches—that is, if they floated when tossed in—they were presumed guilty and might be hanged or burned. If they sank, they were considered innocent, but they might have to drown to prove the point. A bold criminal who knew he was a sinker and was good at holding his breath might rob and pilfer with impunity, safe in the assurance that the waters would absolve him.

When a person drowns, anaerobic bacteria in the gut and any infected wounds get to work. As they digest sugars and proteins in the tissues, they excrete carbon dioxide and sulfur dioxide. These gases inflate certain body parts, mainly the face, abdomen, and male genitals. It typically takes one to two weeks for enough gas to build up, although many factors—depth, water temperature, the amount of sunlight reaching the body, and the speed at which crabs and other scavengers peel the flesh from it—all affect the pace of decomposition. Once inflated, the corpse breaks free of the suction exerted by silt and mud and rises to the surface, unless a log or other obstruction blocks it. At that point, it may be scarcely recognizable. Immersion brings a "horrible, lifeless uniformity," in the words of the poet Oliver Bernard.

If the water is cold enough to prevent decomposition, a body may never rise. But once gasifying begins, it proceeds like a juggernaut. I once studied a body whose killers thought they'd sunk it for good with sections of brick

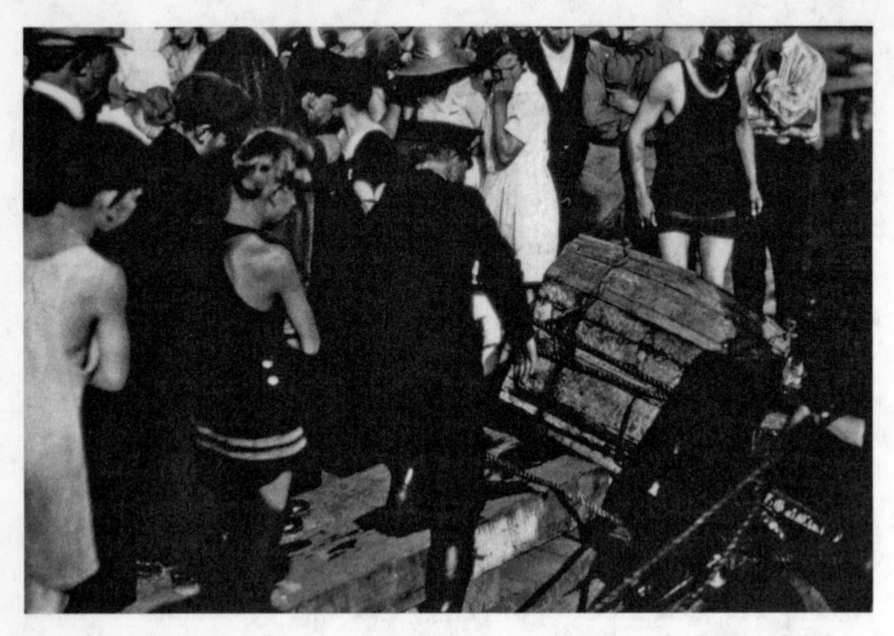

The trunk containing Mahoney's victim is taken from Lake Union, after bacteria foiled his plans.

wall. They had tied five mortared bricks on the chest and five on the back. It still rose to the surface. In April 1921, a recently paroled highway robber named James E. Mahoney learned a lesson in bacterial buoyancy, to his great regret. The thirty-eight-year-old Mahoney had married a wealthy septuagenarian Seattle widow. Ten weeks later he became incensed when he discovered she'd severed him from her will. He poisoned her, wedged her body into a steamer trunk, and tried to dispose of it in Lake Union, just north of the city center. Alas, the trunk floated. He added chunks of concrete and watched, relieved, as the trunk sank. Thinking he was home free, Mahoney forged his wife's signature to get at her funds. But he hadn't done his homework. He didn't realize how shallow the lake was. As spring turned to summer, its bottom waters warmed. Four months after he sank it, the trunk surfaced, buoyed by the putrefying corpse. The next year, Mahoney hanged.

So inexorable is the putrefaction process that if a body is trapped by, say, a leg, the gases may build up until it is buoyant enough to detach itself from the limb and float to the surface. Then you have a missing-foot mystery—the opposite of the recent Canadian cases.

Sharks can detach limbs much faster than bacteria or crabs. They can also store food intact in their stomachs for three weeks or more, digesting it as needed. This is not only a useful knack for a predator that must cruise sparse waters seeking elusive prey but a surprising source of forensic evidence. Tally Calvert once told me about a surfer who disappeared off Galveston in the 1970s. Six days later his board turned up without him. A month later, fishermen happened to find a foot in the belly of a shark. A missing toe confirmed that it was his.

Such evidence may explain a disappearance, but it wasn't enough to clinch a murder case. In 1935, a fourteen-foot tiger shark that had gotten tangled in fishing gear went on display in Sydney's Coogee Aquarium. Within a week it became ill and vomited up a muttonbird, a rat, and a muscular human arm. Fingerprints and a tattoo depicting two boxers proved that the arm belonged to a former boxer named James Smith who'd disappeared nearly three weeks earlier. The police established that Smith and several co-conspirators had wrecked a yacht for the insurance money. When the scam soured, his comrades murdered him and stuffed his body into a tin box. When it proved too small to contain his boxer's bulk, they severed his left arm and tied it to a boat anchor, where the shark found it.

The rest of Smith's body was never found. And despite strong circumstantial evidence, the Australian Supreme Court, citing an English case from 1276, ruled that a single limb was not proof of murder; a one-armed James Smith might yet be alive. The case against his accomplices failed.

Perhaps another shark got the rest of Smith. Large fishes have been known to swallow men whole. Occasionally those swallowed live to tell the tale. One unconfirmed (and rather incredible) account has it that a giant grouper downed a diver off Hawaii, but he wriggled out through its gill. In 1758, a twenty-foot shark swallowed a sailor who'd fallen overboard. His captain immediately harpooned it and he lived to tour Europe, exhibiting the dried shark. In 1771, a sperm whale smashed a skiff, swallowed a harpooner named Jenkins, and moments later spat him out unhurt. It's possible a whale or shark really did swallow Jonah, though he wouldn't have survived long enough in its belly to travel far.

Not even Jonah made so epic a voyage as the nameless transoceanic floater that the forensic dentists called Barnacle Bill, after the barnacles growing on his skull. I first learned of him from another Bill, medical examiner Bill Haglund. Over dinner at our neighborhood New Mexican restaurant, Bill casually mentioned an unidentified skeleton that had washed ashore in Hawaii in a survival suit—standard gear, designed to keep capsized fishermen dry and warm on perilous northern seas, but not in Hawaiian waters. I eventually visited Dr. Charles Odom, Bill's counterpart in Honolulu, confirmed the story, and examined his office's files.

One morning in 1982, a jogger on Punaluu Beach stumbled upon a

Reenactment of the skeleton discovered in a survival suit.

Gumby-like orange body suit. A skull, bleached white and stained with algae, stared from the face hole, barnacles in its eye sockets. At the morgue, Odom's staff sliced open the barnacle-infested suit to find decomposed organs still clinging to barnacle-infested bones. To Odom's trained eye, the skeletal features suggested a thirty-something Caucasian male, up to six feet tall. The fillings on his molars appeared rather primitive—perhaps done in Russia or Asia. He wore only blue jeans. His left forearm was missing but his suit was intact, making it unlikely that any bones had escaped; he'd had only one arm before he went overboard. The examiners estimated he'd drifted for about two years.

I've investigated Barnacle Bill's case since, but I never managed to identify him. I suspect he drifted from the Arctic Ocean; two years would be a rapid but conceivable journey along this path, as confirmed by drift cards that have traveled from the MacKenzie River Delta just east of Alaska's North Slope to San Diego. In April 2008, more than two decades after he washed up, I received a moving message from a young woman in Tacoma who was still haunted by a newspaper article she'd seen mentioning him five years earlier. "It's come to me through the years," she wrote, "and it is to the point that I just have to ask you." Her father disappeared during a scuba lesson in the surging Tacoma Narrows in the early 1970s, and his body was never recovered. Could it have been the skeleton washed up in Hawaii?

Assuredly not, for several reasons. Her father was African American, not Caucasian, and he disappeared a decade earlier, in scuba gear rather than a survival suit. But given the local currents, I told her, "Your father may have drifted very far in a short interval. I hope this helps a bit."

The ancients read much into not just whether but how cadavers floated. "The dead bodies of men float upon the back," wrote Pliny, "those of women with the face downwards; as if, even after death, nature were desirous of sparing their modesty." But George Parsonage of the Glasgow Humane Society told me that Pliny had it exactly wrong: "While most corpses rise with spine uppermost, women and obese persons may rise face-up. This is caused by gas forming in breasts and large abdomens." But, he noted, "contrary to common belief, there is no significant difference between the time it takes for male and female corpses to resurface."

The buoyant afterlife of corpses means that sea burials aren't as easy as they sound. Take the posthumous voyage of Ted Krol of Eugene, Oregon, who asked to be buried at sea. Krol's friends studied state rules requiring that such burials be conducted at least three nautical miles from shore and in water at least six hundred feet deep. They built a cedar coffin, engraved it with a cross, drilled it with holes so it would sink, and conducted the ceremony a suitable distance off Florence, Oregon. Two days later, at dawn, a man walking his dog found Krol washed up. He'd drifted twenty-six miles in forty hours. If his friends had dispatched him farther out to sea, he might have escaped the coastal currents and drifted across the Pacific.

To insure that those buried at sea stayed buried, the navies in the days of sail devised a fail-safe formula, which Young E. Allison evoked in his oft-sung verse "Derelict":

> *Fifteen men on the Dead Man's Chest—*
> *Yo-ho-ho and a bottle of rum!*
> *We wrapped 'em all in a mains'l tight,*
> *With twice ten turns of a hawser's bight,*
> *And we heaved 'em over and out of sight—*
> *Yo-ho-ho and a bottle of rum!*

A ship's sailmaker would stitch canvas around the deceased, inserting twelve pounds of sand, rock, shot, or whatever was handy, heavy, and expendable. Scientific buoyancy tests performed with ninety-eight adult males have confirmed the sailmaker's rule of thumb: an extra dozen pounds sank every body with four pounds to spare.

But sometimes crews skimped on the ballast. In 1799, during the Napoleonic Wars, Commodore Francesco Caracciolo turned against Naples's King Ferdinand IV. Admiral Horatio Nelson, Ferdinand's British ally, captured Caracciolo and had him tried and hung for treason. But the burial detail misjudged Caracciolo's buoyancy and the corpse bobbed to the surface, head above water. The superstitious Ferdinand took fright, but Sir William Hamilton, Britain's ambassador to Naples, suggested that Caracciolo had merely returned to beg forgiveness.

A few mariners are fortunate enough to be spat up by the sea while they are still alive. Around 1942, U.S. Coast Guard Commander Robert W. Goehring was swept overboard off the cutter U.S.S. *Duane* by a monstrous wave. As he neared the ship, another big wave lifted him to deck level, and his crewmates were able to dispense with rescue gear and pluck him aboard.

I don't know how Commander Goehring felt after his lucky escape. But there comes a time when those who work at sea stop believing they will be lucky the next time. Ten years after Nelson Eddy hit, I found myself underway on the Gulf, running tracks around an oil rig and measuring the passing currents with the latest high-tech meter, an acoustic Doppler current profiler. I drew the current patterns and let the rig's captain know when he was likely to encounter a current so powerful—two knots or more—that he would have to shut down his operation to avoid potential catastrophes, which would idle three hundred workers at a cost of about $200,000 a day.

This was the sort of real-time, on-site oceanography I relished, like chasing snarks around Dabob Bay or water slabs around the mid-Atlantic. Indeed, when I worked at Mobil I fought for years to get permission to visit an offshore platform and see firsthand what I'd been designing for. Now I drove to the heliport in Louisiana and rode a chopper out to the giant rig. There I was lowered in the "basket"—a euphemism for a sort of giant macramé assemblage that you cling to while the crane operator lowers you seven stories. But somewhere along the way I had developed acrophobia; I feared I would let go, with no harness or safety net, and asked the captain to have someone ride with me. All went well and the crane operator set me down gently on the rolling research vessel. But I realized I was getting too old to swing around offshore rigs. Soon after that cruise, I phoned Bob Hamilton and told him my offshore days were over. He was very understanding.

6. The Admiral of the Floating World

And much of earth and all the watery plain
Of ocean call'd him king: through many a storm
His isles had floated on the abyss of time. . . .

—Lord Byron, "The Vision of Judgment"

Beachcombers often tell me they live too far from the open sea to find anything interesting. I tell them that wherever the tides reach, ocean drifters will arrive. Each flood tide ratchets flotsam inland, till it strands. The next flood carries it farther, and on and on. Ocean flotsam even reaches us here in Seattle, 120 nautical miles up the Strait of Juan de Fuca and Puget Sound from the Pacific. One drift card from Alaska's Bristol Bay passed Seattle and reached Fox Island, below the Tacoma Narrows Bridge. I learned of it when Vern Larson, the father of the cartoonist Gary Larson, called to tell me he'd found a drift card I'd released in Puget Sound for a sewage outfall study. I asked what other flotsam he'd found—always a good question to remember to raise. Well, he said, there was this card from the Bering Sea. . . .

As on the Pacific, so on the Atlantic and its inland catch basins. Atlantic flotsam penetrates hundreds of miles into the Baltic Sea, two thousand nautical miles into the Gulf of Mexico, and fifteen hundred or more into the Mediterranean. More than two thousand years ago, the Greek philosopher Aristotle recognized this. "In the narrow passage of the mouth, the

Pillars of Heracles, the Atlantic Ocean produces a current flowing into the inner sea as into a harbor."

The cause actually lies at the sea's other end. The dry, intense heat of the Middle East causes so much water to evaporate from the eastern Mediterranean that the equivalent of five Amazon rivers rushes through the Straits of Gibraltar to fill the deficit. For flotsam borne into this sea, there is no escape.

In summer 1999, George Wood, who was vacationing near Malaga on Spain's Costa del Sol, panicked when he spotted what he thought was a human baby washed up on the beach. It proved to be a ten-inch doll strapped to a log with a plastic bottle containing a note: "We can't afford to look after this baby. Please give her a good home." The prank had been hatched by two schoolchildren who'd found the doll on a Scottish beach two months earlier and sent it speeding along the Atlantic currents—at thirty nautical miles a day.

The odds of such a passage are long but not minuscule; only 2 percent of bottles launched from the eastern United States are found and reported in Europe, and I calculate that 1 percent of flotsam reaching Europe travels into the Mediterranean. Crunching these figures, I'd guess that only one message bottle in five thousand from the East Coast reaches the Mediterranean. I know of five other message bottles that did. One, a "Bacardigram" that the yachtswoman Kay Gibson released off North Carolina, was discovered a year later at Genoa, the birthplace of Christopher Columbus, the first flotsamologist to read the messages borne across the Atlantic.

I awakened to Columbus in 1996. That July my father died, after years of struggle. Drained and saddened, I realized I needed to get away. A chance beckoned: A Florida beachcomber named Cathie Katz, the celebrated "Sea-Bean Lady of Florida," had invited me to attend the first Sea-Bean Symposium, a beachcombers' fair she was assembling in Cocoa Beach.

Cathie wanted to hold the fair in October, when the most flotsam washes up along eastern Florida. As it happens, that's also when Columbus first landed in the Americas. It was there, sitting before the public with all the assembled sea-bean experts in the Cocoa Beach Public Library, that I first learned of the remarkable propagative strategy of the sea-hopping seeds called "sea beans," of their deep roots in history and lore, and of the influence they exerted on the first mariners to cross the Atlantic—both the Vikings and Columbus himself.

Every ocean has its floating jewels. Along the North Pacific, the beachcombers' favorite prize is glass fishing floats. Along the North Atlantic, sea beans have been the prize for centuries. In 1570, the pioneering French botanists Mathias de L'Obel and his student Pierre Pena noted that they had "received as a gift from that most distinguished lady, Dame Catherine Killigrew . . . many other very rare beans which are said to be found in great plenty on the shores of Cornwall . . . Year by year they find fresh beans, some floating, others of them digged up from where they lay buried in the sand by the shore, as if they had been drifted from the New World by favoring southerly or westerly winds, as is the faith of the Cornish folk that dwell by the English sea." Artisans fashioned snuff, match, and tinder boxes from these elegant bits of flotsam.

The term "sea bean" embraces a wide variety of seaworthy seeds that propagate by riding to new lands on the Johnny Appleseed currents. They range in size from tiny moonflower kernels to the watermelon-sized coco de mer, unique to the Seychelles, whose haunchlike double lobes titillated the pining sailors of yore. But it most often refers to the hard, smooth-shelled, almond-sized seeds released by various trees and vines that grow along the rivers of tropical America.

Many species can float for years, buoyed by material density and air pockets inside their impervious shells. In 1974, sea-bean guru John V. Dennis set a variety of seeds afloat in an experiment that continues to this day. Seven species, including the nickernut (*Caesalpinia bonduc*), huskless coconut (*Cocos nucifera*), sea coconut (*Manicaria saccifera*), and sea heart (*Entada gigas*), are still floating. Such endurance may help sea-bean species outrun global warming. Wherever they find frost-free habitats they will seed new jungles, spreading northward as the frost disappears.

Beauty and drift go together; the most beautiful sea-bean varieties seem to float longest. Good luck may also ride the currents: peoples around the

Three celebrated sea-bean varieties: hamburger bean, sea heart, Mary's bean.

world consider sea beans fortunate talismans, and the old-time sailors believed they assured safe passage across the sea. The Norse and the Hebrides islanders believed that the seed called "Mary's bean," which bears a cross on one side and a dark outline shaped like an infant in the womb on the other, assured a successful pregnancy. It's a charming notion, but in the only modern test I know of it didn't pan out. Ed Perry, who inherited the flotation experiment after John Dennis's death, offered his wife a Mary's bean to dull the labor pains when she gave birth. It didn't work, and she hurled the bean back at him.

Many in the tropics believed that a sea bean worn around the neck could ward off the evil eye. The native Karankawas of the Gulf Coast esteemed what's now called the "hamburger bean," after the dark, pattylike band around its rim. Álvar Núñez Cabeza de Vaca, the marooned Spanish explorer who lived as a slave and traveled among the Gulf and Southwest tribes in the early 1500s, discovered that they held this "fruit like a bean of the highest value [to] use as a medicine and employ in their dances and festivities." So he did a profitable trade in the beans, traveling among the interior tribes as an itinerant peddler. Having beachcombed Matagorda Island near the spot where Cabeza de Vaca was marooned, I can testify that he could have speedily collected more beans than he could carry. And so, like any experienced beachcomber, he would have selected only the best or rarest specimens.

Once, when I lectured in Newport Beach, California, a man from Cuba recognized a hamburger bean I'd shown and said it was still considered very good luck there. Jean Andrews, author of *Shells and Shores of Texas,* has found hamburger beans "in the herb sections of Indian markets throughout Mexico, where its magical powers are still respected." That respect is grounded in science: hamburger beans contain the dopamine precursor L-dopa, which is used to treat Parkinson's disease. Shortly before my dad died, I told him this and he laughed, wondering if sucking on a hamburger bean would have helped him fight Parkinson's disease.

Shape determines how an object moves in the sea and how it makes landfall, so the odd shapes of sea beans may be evolutionary adaptations that help them wash up where they can best survive. Call it survival of the shapeliest. Surely one of the most shapely is the seed commonly known as the "sea heart." Somehow, evolution and the sea have fashioned an oceanic valentine, a heart-shaped seed that can float for at least thirty-four years. Something in nature seems to love that shape; surf and shore occasionally grind ceramic shards into hearts. C. S. Lambert and Pat Hanbery show fifteen heart-shaped shards in their book *Sea Glass Chronicles*. "Sometimes the sea spins a peculiar kind of magic," they note.

At Cocoa Beach, I also learned that in the Azores the sea heart is called *fava de Colom*, "Columbus's bean," because, as Dennis and Perry put it, "The people think that these are what gave Columbus one of his best clues to the Americas."

Intrigued by this allusion, I read and reread Samuel Eliot Morison's magisterial biography of Columbus, and I noticed an aspect of his quest that is commonly overlooked: the vital role of flotsam. I turned to other biographies, particularly the one written by his son Ferdinand, and the biographies of the other great explorers of his era. And I realized how much he had in common with today's beachcombers. They see, really see, what's carried on the water, and so did he. Other explorers did not—and he, not they, sailed to a new world. He was the beachcomber who changed the course of history.

Columbus had ample opportunity to see American flotsam. As the son of a Genoese weaver, he grew up far from the Atlantic gyre that became central to his life (and which, if this book has its way, will bear his name). But the Mediterranean is a catch basin, and even as a child he might have found mysterious hard seeds as he played along the shore. Consider the numbers. The intact tropical forest of yore released perhaps a billion sea beans a year. Assume that these crossed the Atlantic and drifted into the Mediterranean at the same rate as the drift bottles released in various experiments, about one in five thousand. Divide the result by the number of nautical miles of Mediterranean shoreline, twenty-eight thousand. An average of something on the order of seven sea beans would wash up on each mile of Mediterranean coast each year.

Then Columbus went to sea. He worked for twenty-some years as a ship's officer, sailing south to Africa and north as far as Iceland. He learned which winds blew west and east at which latitudes. Perhaps he heard how the Icelandic settlers used flotsam to find new harbors. Certainly he drank in the taverns of many ports, getting to know sailors from many lands and hearing their tales of strange things seen on the water. Perhaps he saw exotic flotsam in the cabinets of curiosities that were then the rage. Once he became flotsam himself, when he was shipwrecked in a sea battle and, clinging to an oar, washed ashore on the doorstep of Portugal's Prince Henry the Navigator, founder of the world's first oceanographic school.

And, perhaps most important, he spent ample time in the Azores, the prime dumping ground for the great garbage patch around which the North Atlantic Subtropical (Columbus) Gyre orbits. Islands everywhere become magnets for flotsam as currents swirl about them. But a high-pressure atmospheric cell above the Azores creates a veritable vortex of wind and current around them, making them supermagnets—or rather, a great vacuum cleaner sucking up the sea's dust bunnies. In study after study, these geographic flyspecks have drawn an outsize share of Atlantic drifters: 16 percent of the floats recovered from Prince Albert I's tests in the 1880s, 10 percent of 480 transatlantic crossings tabulated by Dean Bumpus, 16 percent of the 639 that John Dennis tallied to gauge the transatlantic drift of sea beans. For every American message bottle that makes it to the Mediterranean, around one hundred wind up on Europe's Atlantic coast and a thousand wind up in the Azores.

The Gulf Stream shakes like a loose fire hose from its pivot point at North Carolina's Cape Hatteras, spraying uncountable drifting objects east toward Europe. Columbus saw the results washed up on the Azores: sea hearts, timber bamboo, driftwood, even derelict kayaks. The driftwood was so fresh, he reasoned, that it could only have crossed a narrow ocean. Westward of the Azores a Portuguese ship's pilot had picked up what Morison calls "a piece of wood ingeniously wrought, but not with iron." On Porto Santo Island in the Azores, Columbus's brother-in-law Pedro Correa da Cunha recovered a similar piece of carved driftwood, as well as thick canes (timber bamboo). His son Ferdinand recorded that "some persons in the Azores also told Columbus that after the wind had blown for a long time from the west, the sea cast on the shores of those islands (especially of Graciosa and Fayal) pine trees that do not grow on those islands or anywhere in that region." Everywhere he saw driftwood that was too fresh to have traveled many thousands of miles.

"Men of Cathay which is toward the Orient have come hither," Columbus himself wrote in the margin of his copy of Aeneas Sylvius's *Historia Rerum*. "We have seen many remarkable things, especially in Galway of Ireland, a man and a woman of extraordinary appearance in two boats adrift." Morison believed, implausibly, that these "fin-men" or "Finnmen," as they were called, were probably Finns or Lapps. But neither people were kayakers, and the currents in the Baltic Sea and off the Norwegian coast would drive any such boats to the east and north rather than southwest into the Atlantic.

I agree with Columbus that the Finnmen would have come from farther west, perhaps Greenland or Labrador. Winter weather patterns show how: Occasionally the jet stream pushes the frigid air mass hanging over North America out over the warm Sargasso Sea. This has a double effect. The strong offshore winds probably pushed many an unfortunate Eskimo out to sea. At the same time, the cold air chills the water below it, increasing its density. The air's dryness also causes evaporation, boosting the remaining water's salt concentration and hence further increasing its density. Together they make the Sargasso surface water so dense it sinks a mile deep. Some kayaks likely sank with the heavier water. But others would have floated on to the Azores and Europe. Protected by seal-gut clothing and the winter chill, corpses could easily have arrived intact enough for Columbus to recognize them as "Chinese."

Some may have even made the passage alive. I found hints to that effect

Native hunter in Greenland ca. 1944, paddling kayak similar to those Columbus thought had drifted from "Cathay."

in a reference by the modern British drift oceanographer John Carruthers to a long-forgotten paper by a nineteenth-century German geographer named Thaddoaus Eduard Gumprecht. Intrigued, I sought out the 1854 German geographic journal in which Gumprecht's work had been published. I located just one copy in the United States, in a South Carolina library, and managed to borrow it through the intercession of a kindly NOAA librarian. The pages had never been cut apart; it appeared that no one in America had ever read Gumprecht's pioneering work. I secured a translator, then discovered that Gumprecht wrote in the now-archaic high German of scholars in his day. It took my translator months to work through it.

Among the sources on early human drifters that Gumprecht uncovered was a reference by Pliny to a lost work from the first century BCE by the Roman historian Cornelius Nepos. It reported that a German prince had presented some "Indians" storm driven onto the coast to Quintus Metellus Celer, the new governor of Gaul. Other arrivals were recorded in the tenth and twelfth centuries on the German coast. In 1508, the earliest testimony clearly identifying the "Finnmen" as Eskimos came from a French warship that captured a small boat off the English coast. The boat contained seven people of "smallish build, darkish skin color, and a generally wonderful appearance, whose language no one spoke and whose clothes were sewn together from fish skins," wrote Gumprecht. "The boat was built of twigs and wood." The sixteenth-century poet and scholar Cardinal Pietro Bembo noted that "their broad faces with the habit of eating raw flesh and drinking blood would imply that they were Eskimos. . . . Six of them died; the sole survivor was taken to Louis XII."

The seventeenth-century Orkneyan physician James Wallace, Jr., reported that Finnmen often washed up in the Orkney Islands. He described one in unmistakably Eskimo terms: "His shirt he has fastened to the boat that no water can come into his boat to do him damage, except when he pleases to untie it, which he never does but to ease nature or when he comes ashore." Others were reported in the Orkneys in 1682 and 1684, riding small "fish skin" boats, and stories have trickled down about a population of Eskimos that once resided in the British Isles.

Arduous though the passage might be, it seems plausible that Eskimo hunters, masters at fishing and seamanship, could survive it. At two sea miles per hour, a reasonable drift speed for such light boats, the twelve hundred miles from Greenland's Cape Farewell to Orkney could take twenty-five days and the fifteen hundred miles from Newfoundland to Ireland thirty-one days. Aleuts paddled up to a thousand miles across the Gulf of Alaska in their kayak-like *baidarkas*—and, lashed on by Russian overseers, proceeded as far as California in pursuit of sea otters. In modern times, people less fit for the ordeal than Aleuts or Eskimos have survived sixty days or more adrift in small boats. In 1928, German-born Franz Rohmer crossed from Portugal to Puerto Rico, nearly four thousand miles, in a canvas kayak. Other kayakers have since traversed the Atlantic by the northern west-east route and crossed the Pacific from Monterey to Maui.

From washed-up clues, Columbus deduced that land lay months—not years—away to the west, within range of his sailors' endurance. He was right; the westernmost Azores lie only about one thousand nautical miles from Newfoundland. He believed that land was the China and Japan he had read about; though he had only slight formal education, he read assiduously, filling a copy of Marco Polo's *Travels* with his notes. He saw that the corpses that washed up in strange little boats were not European, that they seemed to match Polo's description of the Chinese, and he assumed they were just that. Knowing that bamboo did not grow in Africa, he surmised that the pieces he saw came from "Cipangu," Japan. He may have thought he saw the coconuts Polo described in China, even though coconuts, which have since been widely planted, did not then occur anywhere around the Atlantic. Later he reported finding one on northeastern Cuba. What he found was likely the much smaller sea coconut, an entirely different plant that did grow in the Americas, and whose white-fleshed fruit does wash up in the Azores.

The expert geographers and navigators of his time knew, just as the ancients had, that Asia could not be so near; the planet was far too large, and surely flotsam could not have crossed such a vast sea. But Columbus read the waters, not the maps. He trusted the flotsam he could see, not the geometry that had come down from the Greeks. He was an ecological navigator, steering as much by drift and bird sightings as by compass, stars, and sounding lead. His sea logs tell the tale; almost every day he recorded passing drifters—seeds, branches, sargassum weed, shipwreck debris. To his mariner's eye, the sea was marked with signs as a highway is for a modern motorist. He learned to see across an ocean.

Columbus followed the sort of wisdom I used to try to impart to thesis writers: Derive your results five ways, cross-check your observations with various lines of reasoning, then stick with them even if they fly in the faces of the authorities. His assumptions were wrong; he believed the world was much smaller than it actually is. But his deductions were brilliantly perceptive. The floating world led him to the New World.

When he landed, Columbus knew he'd reached his goal by the sea beans and bamboo shards that littered the beach. Combing Florida's Bean Coast, I find the same two types of drifter (they tend to wash up together) and feel myself seeing through Columbus's eyes.

When Columbus finally set forth from the Canary Islands, on September 9, 1492, he merged into one of earth's great oceanic highways. Two other currents, the Gulf Stream to the north and its Pacific counterpart, the Kuroshio, are more renowned as speedy supercurrents. They surge north along the western rims of their respective oceans at a robust hundred miles a day. But the real marathon racers are the east-west currents that feed them—the Pacific Equatorial Current, which runs seven thousand miles from North America to Asia, and the Atlantic Equatorial Current, which hurtles three thousand miles from Africa to the Caribbean—the current that Columbus caught. They average ten miles a day, day in and out, as regular as bus routes.

And so Columbus made good time. He landed at San Salvador in the Bahamas on October 12, after sailing 3,466 nautical miles in 32.8 days. Wind and sail supplied 96 of the 106 miles he averaged each day. But the current carried him the other ten—328 miles in total, and that was the crucial margin of success. As the voyage dragged on, Columbus's crew rebelled, and he promised to turn back if they did not sight land within three days. On the third day they did. But for a current, Columbus would have had to give up, and America might be a very different place today (minus the name "America," for starters). "If the European discovery had been delayed for a century or two," notes Morison—an unlikely prospect, given the pace of exploration then, but *if*—"it is possible that the Aztec in Mexico or the Iroquois in North America would have established strong native states capable of adopting European war tactics and maintaining their independence to this day, as Japan kept her independence from China."

Two days out of the Canaries, Columbus spotted a mast from a ship comparable in size to the *Santa Maria* on which he sailed, floating westward. He took the time to record the sighting in his log because, I believe, it confirmed early in the voyage that he was on the right oceanic road. This was only the first of many artifacts from the Old World that Columbus would encounter in the New World. On his second voyage, a search party he dispatched to the island of Capesterre found what his biographer Morison concludes was a timber from a Portuguese caravel that had floated over from Africa. And on the third voyage, a Beata Islander came aboard one of Columbus's ships carrying a fully equipped European crossbow.

Columbus and his crew inevitably encountered the vast mats of *Sargassum* weed that float, unattached, in the mid-Atlantic area. And they thought they observed another phenomenon whose existence is more controversial, though hundreds of sailors have reported it over the centuries: floating islands.

Ten days west of the Canaries, in the midst of the *Sargassum*, Columbus logged that he was "going through islands." His crew "began to see many tufts of grass which were very green, and appeared to have been quite recently torn from the land," one seaman, Pedro de Velasco, recounted. "From this they judged that they were near some island." Pedro had advised Columbus to steer a straight course through the *Sargassum*, which yielded easily to a ship's prow, rather than trying to follow the twisting paths between the mats—a sure way to get lost. But the wind fell, and the three little ships became becalmed. Fearing they would drift onto the supposed islands, Columbus ordered the leadline lowered six hundred feet but found no bottom.

Certainly he had heard of uncharted islands sighted in the mid-Atlantic; several were reported each year, and these reports had helped inspire his voyage. Sailors would often find themselves approaching islands in the evening, then see no sign of them in the morning, and cartographers duly entered these on their maps. Columbus himself made at least two false landfalls, causing his crewmen's spirits to soar and crash. As a result, he amended his promise of a reward for the first man to spot land. "In order to prevent men from crying 'Land, land!' at every moment and causing unjustified feelings of joy," Columbus's son Ferdinand later wrote, "Columbus ordered that anyone who claimed to have seen land and did not make good his claim in the space of three days would lose the reward even if afterwards he should actually see it."

Morison sneers at what he calls "phantom" islands: "As for the islands reported west of the Azores, only a person who has never been to sea would believe in their existence. Sighting phantom islands and disappearing coasts is a commonplace of ocean voyaging. A line of haze, a cloud on the horizon (especially at sunset) often looks so like an island as to deceive even experienced mariners who know that no land is there. In Columbus's day, when anything was possible, a shipmaster sighting an imaginary island at sunset would set a course for it if the wind served, and when day broke and no land appeared, he would conclude that by some compass or other error he had passed it in the night. . . . If every island were real that some mariner has thought he sighted during the last four centuries, they would be as close together as the Florida Keys."

But might some of these phantoms have been real, even if they were not fit for maps—floating islands, there and gone? The influential early-twentieth-century paleontologist William Diller Matthew estimated that a thousand islands drifted out to sea during the seventeenth, eighteenth, and nineteenth centuries, and 200 million during the Cenozoic era. Such islands, formed when soil collects on dense mats of fallen trees and other debris, were known on the lakes of Europe, the marshes of Mesopotamia, and the log-jammed rivers of the Pacific Northwest. On the Nooksack River, which empties into Bellingham Bay north of Puget Sound, a logjam grew until it stretched for three-quarters of a mile and rose as much as twenty feet above the waterline, with hundred-foot firs growing from its marl. Official surveys in the 1860s revealed two such jams spanning the Skagit River to the south, one a mile-and-a-half long. Its surface, one pioneer recounted, "supported a forest growth scarcely distinguishable from that prevailing on the river's banks. This forest rose and fell with the rise and fall of the river. In times of flood, owing to the settling and shifting of the mass in the upper regions of the jam, a weird note of groaning was produced not unlike that of a monster in pain, while sharp reports of breaking timber could be heard for miles."

On the open sea, such noise would carry even farther. Perhaps it was the grinding of matted logs that made so many sailors hear monsters roaring across the deep and fear floating islands as sinister portents. Today engineers and harbor authorities clear out such accumulations before they block passage and menace shipping. But untended, they would pile up until enormous floods washed them out to sea, there to drift, taunting mariners and bedeviling mapmakers, until they broke apart on the waves or crashed onto new shores.

Floating islands certainly occur on the more sheltered waters of lakes. In the first century CE, Pliny visited Lago di Bassanello, near the Umbrian town of Amelia, and beheld a remarkable sight: "No vessels are suffered to sail here as its waters are held sacred. But several islands swim about it, covered with reeds and rushes, and whatever other plants the neighboring marsh and the borders of the lake produce." Sometimes these islands would cluster in a "little continent," then scatter and seem to race each other across the lake, then attach themselves to the shore. Sheep would step onto them to graze, start at finding themselves surrounded by water, then calmly step off when they reached the other shore.

Cattle grazed on an island that grew upon large oak logs on Scotland's Loch Lomond and broke from shore in 1796. By 1835, large alders and willows had rooted, and William Wordsworth celebrated it as "a mossy islet with trees upon it, shifting about before the wind." A decade later, the raising of a lake at Capesthorne, England, dislodged a two-acre island covered with silver beeches and brush. It blew about the lake for a few years, then lodged on the protected northeast shore and resumed its former status as terra firma.

Across the world, 12,507 feet above sea level in the Andes, the oldest continuously inhabited floating islands in the world drift on the deep blue waters of hundred-mile-long Lake Titicaca. In 1532, Quechua Indians fleeing Pizarro and his fellow conquistadores hid on the floating isles called *uros*. The islanders build uros by joining the floating root clods of the tall cattail known as *totora* and overlaying them with cut reeds, which they also weave into huts and boats. After several months the reeds lose buoyancy and the islanders lay down new totora. Finally the uro is large enough to support a village. In 1994, 150 inhabitants lived on the largest of seventy uros, 13,000-square-foot Tribuna.

A thousand miles to the south in the Chilean Andes, no less an observer than Charles Darwin, riding from Valparaiso to the gold mines of Yaquil, saw islands "composed of the stalks of various dead plants intertwined together, and on the surface of which other living ones take root. . . . As the wind blows, they pass from one side of the lake to the other, and often carry cattle and horses as passengers." Nearly 170 years later, in 2002–3, researcher Toby Ault retraced Darwin's steps and learned that the lake had been drained for farming. So passes the floating world, like the rest of the natural world.

In spring 1892, something much more singular appeared off Florida's east coast. It was a season of extreme weather: hurricanes, tsunamis, and floods violent enough to uproot whole sections of forest. One such section became the only wooded island ever observed traversing an ocean. Thirty-foot trees enabled mariners to see it from seven miles away. The U.S. Hydrographic Office feared it would menace transatlantic steamers, and inscribed it on the monthly pilot charts that marked such threats as icebergs, submarine mines, burning vessels, and floating logs. Many captains stared in disbelief when they received their November 1892 chart for the

North Atlantic; it showed an island floating in the stream. But this was no cloud or mirage; it had been sighted six times along a 2,248-nautical-mile course.

This specter inspired literary imaginations, as magical islands often do. Six years later, Jules Verne published *The Floating Island*, about a man-made isle the size of a very large iceberg. In 1922, in *The Voyages of Dr. Doolittle*, the engineer-turned-author Hugh Lofting described a floating island "drifting southward all the time in a current." Had he too read about the island that drifted halfway across the Atlantic?

I began investigating floating islands shortly before I attended that first Sea-Bean Symposium in 1996. Among the first I uncovered in old periodicals were the islands afloat in Orange Lake, Florida. It so happened that the lake, just four hours north of Cocoa Beach, was where Marjorie Kinnan Rawlings had lived and written *The Yearling* and *Cross Creek*. The creek she made famous drains into Orange Lake.

Islands float through American Indian myth. A thousand miles to the north, the Iroquois imagined a great island floating in space, far from all pain, want, and sorrow, whose ruler was responsible for the earth. Florida's Seminoles interred their loved ones on Orange Lake's floating islands, believing that when the islands sank their souls would rise to the Land of the Sure. I determined to reach those islands myself.

As soon as I'd made my goodbyes at the symposium, I drove north. Hours later, in a howling downpour, I pulled off Highway 75 at Micanopy and ducked into a jiffy store called the Cross Creek Outpost. I waited out the squall sipping jiffy-shop coffee with the Cross Creekers and puzzling out the local lingo. "Oh yeah, floating islands," said one named Billy Carter. "You mean tussocks." I asked him about the spatterdock, also known as cow lilies or *Nuphar advena*, which, I'd read, grow out from shore, forming buoyant mats as large as an acre that are held afloat by gas-filled pads and roots. Strong winds dislodge the mats and the untethered roots eventually rot away, but other symbiotic plants sprout and keep the colonies afloat.

The term "spatterdock" momentarily stumped Billy. "Oh, you mean *bonnets!*" he exclaimed, and I learned another local term. He invited me to come see them the next day—and to join a Sunday outing onto the lake, on air boats. For the rest of that Saturday, another Cross Creeker named Ronnie Thomas shepherded me around the lake. His family once owned a fishing camp and he'd often slept aboard tussocks. Cottontails infested them, he explained. One day he and his dog, Clyde, caught 125 on a single island. Another tussock once floated up to Billy Carter's dock and a horde of rabbits, fleeing two rattlesnakes, bounded off and overran his place.

Sunday morning, coffee in hand, I wandered over to the white clapboard farmhouse at Rawlings's Cross Creek Farm, now preserved as a state park. It wasn't yet open, but a docent invited me in while she tidied up. Tussocks, she said, helped moonshiners hide from revenuers because they constantly shifted and changed the map of the lake. In the 1930s and 40s, Rawlings's neighbor Don McKay conducted nature tours of the tussocks. He'd seen the winds push them three-plus miles a day. One roamed the lake for ten years. Another time, a tussock blew away from a fishing camp, covered three-quarters of the lake's sixteen-mile length, and then, when the wind shifted, glided back to camp. McKay once tried to clear out the creek's mouth with dynamite, but the blast dislodged nearby mats and he had to whipsaw through a hundred yards of newly risen tussocks to escape.

As we taxied on Billy's shiny black air boat, I noticed how the masses of lily roots resembled kelp paddies. The peat on the lake bottom was the color of oversteeped tea. Once, Billy told me, there were as many tussocks on the lake as clouds above it. Today they were few and tattered. Deliberate intervention was partly to blame; in 1998, to open up more space for anglers, the Florida Game and Fresh Water Fish Commission removed acres of tussocks. But an accidental assault had done even more damage. In the 1950s, some careless aquarist had discarded a few sprigs of hydrilla, a water plant native to Southeast Asia and Oceania and imported for home aquariums. A decade later, hydrilla choked Florida's waterways. The weed choking Orange Lake was the same one that sheltered guppies in my childhood aquarium. By the 1990s, the hydrilla was so out of control officials began poisoning it. The currents carried the herbicide around the lake, killing the legendary bonnets. Once again, floating islands got in man's way and were swept aside.

Columbus enjoyed astonishing weather luck on the first leg of his first voyage. He crossed the North Atlantic during the peak storm season but encountered no severe cyclones or hurricanes. Then, on his return trip, he climbed far to the north to find the easterlies that would carry him home and stopped at the Azores, a familiar refueling station. There the auspicious winds deserted him.

Once or twice each winter, the calm Azorean waters are whipped by intense vortices generated by the clash of warm, moist air blowing up from the tropics and cold, dry air down from the Arctic. The winter of 1492–93 was particularly brutal. Cold gusts out of the west pinned ships down for months in the port of Lisbon. Even inside the Mediterranean, Genoa's harbor had frozen up by Christmas.

Columbus sailed the *Niña* into a violent frontal storm, one of the century's most severe. He feared that if he died at sea his rivals the Pinzón brothers, who sailed aboard the *Pinta*, would reap the glory and rewards for his discoveries—and his sons, left unprotected, would go to paupers' prison. So he launched what were, in effect, the world's first confirmed message bottles: He scribbled two summaries of his voyage, sealed them, addressed them to Queen Isabella and King Ferdinand, and encased them in wax, together with the promise of a thousand-ducat reward—some $200,000 in today's money—to the finder who would convey the document to the queen. He secreted these packets in wine barrels and tossed one barrel overboard, hoping it would drift to Europe; his crewmen assumed this was "an act of devotion." The other he set on the *Niña*'s stern so that, as he later recounted, "if the ship sank, it might float on the waves at the mercy of the storm." Perhaps he took the idea from tales he'd heard in Iceland of thunder gods and floating seat posts. Or perhaps, in his reading about Asia, he'd come across Yasuyori's stupa.

But where did the tossed barrel go? The Azores currents disperse flotsam both toward Spain and Portugal (as Columbus hoped) and, more often, to America, along the subtropical Columbus Gyre. The gyre retains about half its flotsam through each revolution; thus there is a 1 percent chance that a given object would remain adrift after seven orbits—twenty-three years. Long before then, however, wood-devouring teredo worms, the termites of the sea, would have bored through the staves.

There's a tiny chance that Columbus's barrel still survives in the Arctic, whose cold waters would stop the teredos. If it happened to turn north along Florida, the Gulf Stream might have carried it to France or Great Britain. If it did not beach there, it could have drifted along Norway to the Arctic Ocean. Wood stranded there has been dated as far back as ninety-five hundred years.

If Columbus's barrel did drift to the West Indies, as I think likely happened given the prevailing winds and currents, his promise of reward would have been meaningless. All around the world, wherever iron was not forged, scraps of it were dearly prized. Using stones, those who found them could shape the metal into valuable tools and weapons. Pacific islanders considered spikes embedded in driftwood more precious than gold. When Bodega y Quadra, the first Spanish explorer to anchor off today's Washington state, sent a party ashore to gather wood and water, he watched as the native Quinault slaughtered his men for the iron in their longboat.

Any American natives who found Columbus's barrel might already have learned to gather iron from previous flotsam. As Columbus's own observations show, wood from Europe drifted readily along the equatorial current. And the Icelandic sagas recount that when the Vikings explored the North American coast five centuries earlier, they discovered the keel of a wreck.

Europeans and Middle Easterners had been filling the Atlantic with ships—and, inevitably, littering it with wrecks—for two millennia. The Phoenicians, the first great Mediterranean mariners, called the Atlantic the "weedy sea," suggesting they'd ventured at least as far as the Sargasso Sea. Punic amphorae have been discovered off the Caribbean coast of Honduras. The Gauls sailed regularly between today's France and Britain and, as their nemesis Julius Caesar conceded, "excel[led] the rest in the theory and practice of navigation." In 56 BCE the Roman navy, with seventy or eighty ships measuring eighty-five feet long, battled 220 Gallic vessels in the Bay of Biscay. Of such encounters mighty shipwrecks are made.

The late Willard Bascom, a respected oceanographer and pioneering underwater explorer I worked with in my Mobil days, estimated that some forty thousand ancient ships may have sunk in deep waters, including "at least 15,500 merchant ships in the first millennium B.C. alone." Following the oceanographer's 2 percent rule—1 to 2 percent of drifting objects released on one side of the ocean will reach the other—pieces of hundreds, perhaps thousands, of ancient ships may have reached the Americas. That's a lot of iron nails and chains to whet the early beachcombers' appetites.

So it seems likely those who found Columbus's barrel smashed it for the hoops holding it together, ignoring the worthless parchment inside. In the end, Columbus did not care; his weather luck held. He sailed though the storm's milder southern side; rough as it was, he almost certainly would have foundered had he sailed a few hundred miles to the north, where winds and waves were higher.

Neither of Columbus's barrels was ever reported found. Neither he nor anyone else realized then that fragile glass could last better at sea than robust wood and iron. If Columbus had used a bottle instead, we might yet have a copy of his log.

7. Borne on a Black Current

Always, then, in this flotsam and jetsam of the tide lines, we are reminded that a strange and different world lies offshore.

—Rachel Carson, *The Edge of the Sea*

Columbus's barrel has disappeared, but other storied drifters float forever on the seas of legend and, lately, the Internet, whether or not they ever existed: the drift bottles Aristotle's protégé Theophrastus supposedly tracked across the Mediterranean, Queen Elizabeth I's "royal uncorker," the ghost ship *Octavius* and the *Sydney*'s phantom lifebelt, Daisy Alexander's will in a bottle, and Clyde Pangborn's ocean-hopping plane wheel. (See appendix A for a parsing of the odds that each of these is actually true.)

These tales have spawned legal battles, comics-page yarns, and endless dinner-table diversion. Other transoceanic drifters have had much larger effects. Some scholars and aficionados believe that ancient drifts brought more than just timbers, nails, and other inanimate flotsam to the Americas. They maintain that sailors, fishermen, or passengers occasionally survived the drift and settled in the Americas, injecting new cultural and genetic elements into its native societies. Some, such as the British-born zoologist and amateur epigrapher Barry Fell, go further. They maintain that Old World peoples—the secretive, sea-mastering Phoenicians in particular—actually sailed to the New World to trade and left their shipwrecked traces off shores as widely scattered as Beverly, Massachusetts, and Rio de Janeiro. Unfortunately, the native peoples of the Americas did not leave records of any such early contacts, so the epigraphers rely on inscriptions and other artifacts—often controversial, if not outright fraudulent—supposedly left by the ancient visitors.

It's harder to argue that Asian voyagers likewise visited or traded with America, because distances across the Pacific are so much wider. And no flood of Asian artifacts has been reported in the Americas to match the European claims. Nevertheless, another contingent of scholars makes a compelling case for repeated wash-ups by Japanese castaways over the past six thousand years—sometimes with transformative effect on the native cultures of the Americas. The doyen of this faction is Betty Meggers, an eminent anthropologist at the Smithsonian Institution, who has advanced this inquiry for more than fifty years despite fierce resistance from her colleagues. In 1966, she published an authoritative account in *Scientific American* of how Japanese mariners drifted to Ecuador five thousand years ago. Since then she's uncovered evidence—DNA, viruses that could only have originated in Japan, and pottery techniques found nowhere else—suggesting that ancient Japanese influence also reached Central America, California, Ecuador, and Bolivia.

Well into her eighties, Betty would present her latest research on Japanese diffusion each year at the Pacific Pathways meetings in Sitka. Before the sessions, we and the other Pathways participants would board a boat to remote beaches near Fred's Creek, an hour from Sitka. Between exclamations of delight at the telltale flotsam we discovered, Betty would share more of her findings. She approached the problem as a literal jigsaw puzzle, comparing pottery shards unearthed around the Pacific. The patterns on multiple shards excavated at Valdivia, Ecuador, and on Kyushu, the southernmost of Japan's main islands, matched so well, she posited that a boatload of Japan's indigenous Jomon people made the trip some sixty-three centuries ago. Other discoveries suggest that others first made landfall in California and San Jacinto, Colombia.

The impetus to this migration was one of the great cataclysms of humankind's time on earth. Few places are so prone to natural catastrophe as Japan, an island nation floating at the intersection of three tectonic plates, the Pacific, Eurasian, and Philippine. The slow but violent collision of these three plates produces spectacular earthquakes, tsunamis, and eruptions.

About sixty-three hundred years ago, a flyspeck island off southern Kyushu named Kikai exploded with a force that would dwarf all the more famous volcanoes that have since erupted around the world. Kikai weighed in at 7 on the standard volcanic explosivity index (VEI), which runs from 1 to 8, VEI 8 being reserved for the sort of mega-eruptions that cause ice ages and mass extinctions. It ejected twenty-four cubic miles of dirt, rock, and dust into the air, about nine times as much as Krakatoa in 1883, twenty-four times as much as Mount St. Helens in 1980, and forty times as much as the eruption of Vesuvius in AD 79 that destroyed Pompeii and Herculaneum.

The tsunamis triggered by Kikai obliterated coastal towns. The eruption's spew was enough to blanket up to 18 million square miles of land and sea. Dust and ash several feet thick smothered the fertile soil, rendering southern Japan uninhabitable for two centuries. Unable to farm, the Jomon set out for other shores in what Betty Meggers calls "the Jomon Exodus." And that was where a second mighty phenomenon came into play.

The Kuroshio ("Black Current," named after the dark color it lends the horizon when viewed from the shore) is the Pacific Ocean's answer to the Atlantic's Gulf Stream. More than twenty-two hundred years ago the Chinese called the Kuroshio by the prescient name Wei-Lu, the current to "a world in the east from which no man has ever returned." Surging up from Taiwan, fat with warm tropic water, it arcs past Japan and Southeast Alaska and down the Northwest coast. At the same time, cool, powerful offshore winds, the equivalent of Atlantic America's Arctic blasts, race down from Siberia, pushing boats and other flotsam out into the Kuroshio.

The fleeing Jomon were driven into the Kuroshio. So were fishermen blocked from returning home by the sea-blanketing pumice. The Black Current bore them toward America—surely not the first and far from the last unwitting emissaries to make that journey.

Europeans call drifting ships "derelicts" once their crews have taken to the longboats. But the Japanese use the word *hyôryû* for a marine mishap in which a vessel, the *hyôryû-sen,* loses control and drifts without command. Traditionally its crew and passengers—*hyôryû-min,* drifting people—would stay aboard, awaiting their fate.

In half of known hyôryû cases, at least some hyôryû-min survived to reach land. And some of those survivors dramatically affected the societies they beached upon. Around 1260 CE, a junk drifted nearly to North America, until the California Current caught it and sent it into the westbound trade winds, which deposited it near Wailuku, Maui. Six centuries later the oral history of the event had passed down to King David Kalakaua, Hawaii's last reigning monarch. As the tale came down, Wakalana, the reigning chief of Maui's windward side, rescued the five hyôryû-min still alive on the junk, three men and two women. One, the captain, escaped the wreck wearing his sword; hence the incident has come to be known as the tale of the iron knife. The five castaways were treated like royalty; one of the women married Wakalana himself and launched extensive family lines on Maui and Oahu.

That was just the first accidental Japanese mission to Hawaii. By 1650, according to John Stokes, curator of Honolulu's Bishop Museum, four more vessels had washed up, "their crews marrying into the Hawaiian aristocracy, leaving their imprint on the cultural development of the islands. . . . Hawaiian native culture, while basically Polynesian, included many features not found elsewhere in Polynesia."

The Japanese presence in Hawaii may go back much further. Hawaiian legend recounts that the first Polynesian settlers there encountered diminutive *menehune* ("little people"), marvelous craftsmen who still dwell in deep forests and secret valleys. At that time, the Japanese were more than a foot shorter than average Polynesians and adept at many strange technologies—from firing pottery and spinning silk to forging metal—that might indeed have seemed like marvels.

Japanese influence likewise spread in mainland North America. Archaeological digs occasionally unearth traces: iron (which native Americans did not smelt) discovered in a village buried by an ancient mudslide near Lake Ozette, Washington; arrowheads hewn from Asian pottery discovered on Oregon's coast; and, of course, the six-thousand-year-old Japanese pottery shards in Ecuador. Just as Betty Meggers found unique artifacts, viruses, and DNA markers in Ecuadoran subjects, the anthropologist Nancy Yaw Davis found telltale Japanese traits in the Zuni of northern New Mexico, distinct from all the other Pueblo peoples. Davis concluded that Japanese had landed in California in the fourteenth century, trekked inland, and helped found the Zuni Nation.

All told, the University of Washington anthropologist George Quimby estimated, between 500 and 1750 CE some 187 junks drifted from Japan to the Americas. The number of drifts increased dramatically after 1603—thanks, ironically, to the efforts of a xenophobic regime to keep foreign influences out of Japan and the Japanese in. In that year the Togugawa shogun, who had united the nation after years of civil war, closed Japan to the outside world, exempting only restricted trade through the port of Nagasaki. Western ships and castaways were to be repelled. Missionaries and other foreigners who entered were to be killed—as were Japanese who left and tried to return.

To ensure that Japanese mariners remained in coastal waters, the shoguns dictated that their boats have large rudders, designed to snap in high seas. Vessels blown offshore were helpless; to avoid capsizing, crews would cut down their main masts and drift, rudderless and unrigged, across the ocean.

Politics conspired with geography, weather, and ocean currents to set this slow-motion, accidental armada adrift. Over the centuries, the shoguns transferred their power to Edo, now Tokyo, and demanded annual tributes of rice and other goods. But Japan's mountainous terrain made land transport impossible, so each fall and winter, after the harvest, tribute-laden vessels sailed from Osaka and other cities in the populous south up the outer coast to Edo. To get there, they had to traverse an exposed deepwater reach called Enshu-nada, the infamous Bay of Bad Water. And

OSCURS simulates the drifts of 1,187 hypothetical drifters released off Japan's Great Cape every ten days from 1967 through 1998. Their paths and destinations vary according to the weather prevailing after each release.

they had to cross just when the storms blew down from Siberia—the same weather pattern that rakes Labrador, Newfoundland, and New England and drives kayaks across the Atlantic. Of ninety drifting vessels documented by the Japanese expert Arakawa Hidetoshi, storms blew 68 percent out into the Black Current during the four months from October to January.

To see where the hyôryû-min drifted, the girls of the Natural Science Club in Choshi, Japan, threw 750 bottles into the Kuroshio in October 1984 and 1985. By 1998, beachcombers had recovered 49: 7 along North America, 9 in the Hawaiian Islands, 13 in the Philippines, and 16 in the vicinity of Japan—percentages remarkably similar to those of the known hyôryû. A few swung back onto the Russian peninsula of Kamchatka, just north of Japan. Kamchatkans adopted the slang term *dembei* for bobbing castaways, after a Japanese fisherman named Dembei whose junk drifted there in 1697—the first known contact between Japanese and Russians.

A few twentieth-century adventurers have traveled as far in open boats as the hyôryû. In 1991, Gerard d'Aboville rowed a twenty-six-foot boat solo for 134 days and 6,200 miles, from Japan to North America. In 1970, Vital Alsar and four companions sailed a balsa raft from Ecuador to Australia, covering nearly eighty-six hundred miles in six months. And in 1952, Dr. Alain Bombard set out to prove that humans could survive being lost at sea by drifting for sixty-five days across the Atlantic in a collapsible raft, catching fish and sipping seawater. But none of these daredevils came near to lasting as long at sea as the hyôryû-min, who often drifted more than 400 and once more than 540 days. Typically just three out of a dozen in a crew would survive—the fittest and most resourceful, who were best equipped to influence, even dominate, the societies they encountered.

As the centuries progressed, the number of Japanese coastal vessels, hence the number of drifters, soared. By the mid-1800s an average of two Japanese derelicts appeared each year along the shipping lanes from California to Hawaii. Four showed up near Hawaii in one thirty-year period in the early nineteenth century; at least five crewmen survived. Many other junks passed unseen along less-traveled routes. During my visits to Sitka, I was afforded the privilege of interviewing many Tlingit elders. I would tell them one sea story, and they would reciprocate with an ancient tale of their own. One elder, Fred Hope, told me that every village along the West Coast has passed down a tale of a Japanese vessel drifting ashore nearby. To the south, around the storm-wracked mouth of the Columbia River, strandings were so frequent that the Chinook Indians developed a special word, *tlohonnipts,* "those who drift ashore," for the new arrivals.

Then, in 1854, a very different landing took place on the other side of the ocean. Commodore Matthew Perry and his "black ships" arrived to open Japan to the world. Perry found skilled interpreters—Japanese who had never left Japan but were fluent in English—waiting to meet him. How could this be in the hermetically sealed hermit shogunate?

The answer lies in the drifts along the Kuroshio. In October 1813, the junk *Tokujo Maru* left Tokyo, returning to Toba after delivering the shogun's annual tribute. The nor'westers swept it out to sea and it drifted for 530 days, passing within a mile of California when offshore winds blew it out to sea. Eleven of the fourteen men aboard perished. Then, 470 miles off Mexico, an American brig hailed the hulk and rescued the three survivors. After four years away, the *Tokujo Maru*'s captain, Jukichi, returned to Japan. Somehow he escaped execution and secretly recorded his travels in *A Captain's Diary*. Though it was officially banned, Jukichi's *Diary* intrigued and influenced Japanese scholars, paving the way for Commodore Perry and for another foreign guest who arrived six years before him. "Unquestionably," James W. Borden, the U.S. Commissioner to Hawaii, remarked in 1860, "the kindness which had been extended to shipwrecked Japanese seamen was among the most powerful reasons which finally led to the opening of that country to foreigners and foreign commerce."

In October 1832, another junk, the *Hojun Maru*, left Toba loaded with rice and ceramics. A typhoon hurled it toward America and, after 450 days, the surf smashed it against the Washington coast, near Cape Flattery. The captain and two boys aboard survived and were captured by the local Makah tribe, to be kept as slaves. When word of the shipwreck reached the Hudson's Bay Company trading post at Fort Vancouver, a party was dispatched to retrieve the three survivors.

At that time, Ranald MacDonald was a boy of ten, the son of a Clatsop princess and a Hudson's Bay fur trader, at Astoria on the other side of the Columbia River. MacDonald never saw the shipwrecked Japanese, but the tale of their accidental crossing inspired him in the same way Eskimo drifters and sea beans inspired Columbus. And he conceived an audacious and, by all lights, suicidal plan: to wash up himself in Japan.

Fourteen years later, at the age of twenty-four, he got the chance to carry it out. MacDonald shipped aboard the whaling ship *Plymouth* and persuaded her captain to set him adrift off a volcanic island near the north end of Hokkaido in a dingy that he promptly scuttled. He was captured but, thanks perhaps to his winning, affable manner, merely jailed rather than executed. MacDonald turned his jail cell into a schoolroom, teaching

English to eager scholars. Six years later, some of his pupils translated delicate diplomatic exchanges between the shogun and Commodore Perry. One of them, Murayama, went on to become the first interpreter in Japan's U.S. Embassy and the first modern Japanese to guide a ship across the Pacific. The drift had come full circle.

Ranald MacDonald joined Columbus, Clyde Pangborn, Theodore Gumprecht, and Amos Wood on my list of drifting heroes—those who circled the ocean in mind or body and revealed its secrets. I have tried to follow where I can in his wake. Susie and I traveled to Fort Vancouver to see the stone marker dedicated to the marooned Japanese sailors who inspired MacDonald, and to Astoria to hunt out the secluded, forgotten memorial to MacDonald himself. Someday I will make it to his grave in Colville, near the Washington-Canada border.

Even after Japan opened up and the shoguns, displaced by the Meiji Restoration, could no longer cripple its junks with fragile rudders, the Kuroshio continued to sweep ships off toward America. In 1876, Charles Wolcott Brooks, the Japanese consul in San Francisco, published a breakthrough paper with a map showing the locations of fifty-five Japanese derelicts. Using these coordinates, his colleague George Davidson sketched the currents of the Turtle Gyre, the first clear depiction to my knowledge of a gyre based on drifters.

Hyôryû-sen have continued to come unmoored right up to the present day. Stored on my basement shelves along with untold other flotsam are emergency rations of water and food, sealed against the sea and inscribed with a message in Japanese, "Don't give up hope. A ship is on the way." So many vessels were swept away after big storms that these packets were tossed like message bottles into the current, in hopes they might reach desperate survivors.

Ships that did happen upon hyôryû-sen or other derelicts could rescue any survivors but not the crafts themselves; they had no way of towing the hulks. And so, when the Russian icebreaker *Taymyr* spotted a drifting junk in 1914, one witness recorded, "We let it drift on its way." He noted that "it had obviously been drifting for some time, since it was entirely outlined in marine specimens." By the time derelicts wash ashore, barnacles and other marine encrustations have usually effaced all evidence of their identities.

That fact makes Rodney Schatz's investigative success all the more notable. Rodney, a stalwart of the Alert Network, lives on northern Graham Island in the Queen Charlottes and frequently beachcombs nearby on magnificent Rose Spit. There he found about 280 washed-up Nike sneakers, out of which he assembled seventy good pairs for his large family. In August 1981, he came upon an eighteen-foot wooden skiff, covered with barnacles, that had just washed up. Rodney scraped it clean, used it awhile, then set it in his front yard and endeavored to discover where it had come from. By corresponding with the Japanese consulate and the Coast Guard he learned that it was the *Hoei Maru*—the "Abundant Treasure"—blown from its moorings off Kamaishi City by a typhoon on March 26, 1980. Now this abundant treasure is his, name and all.

Other derelict passages seem almost predestined. In March 1984, a retired municipal worker named Kazukio Sakamoto pushed out from Owase in Japan's Mie Prefecture for a day's fishing aboard the thirty-four-foot *Kazu Maru*, which even his wife called the "love of his life." He never returned. Eighteen months later the *Kazu Maru* washed up at Prince Rupert, British Columbia, minus Kazukio. Afterward, Prince Rupert and Owase became official sister cities.

And some passages are downright spooky. On Halloween in 1927, the merchant ship *Margaret Dollar* spotted the eighty-five-foot fishing vessel *Ryo Yei Maru* dead in the water off Washington's coast. Human bones littered her deck. Bone-filled kettles sat on a rusty stove. Mummified corpses huddled in one corner. A log scrawled on a cedar board recounted their fate. Eleven months earlier they were fishing off Choshi when a gale drove them eastward, snapping the boat's crankshaft and leaving them to drift on the Black Current. They subsisted on rice, fish, and finally their comrades, until the last man logged a final entry.

Ryo Yei Maru means "good and prosperous ship," but other fishermen believed her cursed by their god Konpira because she flaunted tradition by

Inspectors examine the Ryo Yei Maru *ghost ship in Seattle.*

carrying an engine, ice storage, and five times the usual tonnage. Fearing Konpira, the crewmen's families refused salvage rights, and the cannibal ship was burned without ceremony on a Seattle beach.

Pacific flotsam doesn't just arc eastward from Japan and Asia to North America. It also continues northward and southward along the American coast, from Alaska as far as Ecuador, on journeys of one to two years and two to four thousand nautical miles—drifts that rival transoceanic crossings, though they often progress within sight of land. Since it lacks a conventional name, I call this route the American Coastal Pathway.

According to both OSCURS's calculations and fifty-five recovered research bottles, this pathway saunters along at just seven miles a day—a seventeenth the speed that some rivers course, and just half the distance a sea turtle paddles in a day. Nevertheless, sea turtles follow the pathway to save energy and gain speed as they return to their feeding grounds off Baja California. Derelicts and detritus spin off onto the beaches, like cars abandoned along a freeway: Japanese junks, of course, and the Spanish galleons that sailed from Manila to Acapulco for 250 years. Like the turtles, they followed the current. We know from records and washed-up artifacts that some galleons wrecked off the shores of Hawaii, Oregon, and Baja California, though no actual wrecks have been found. I have been assisting in galleon hunts off Oregon and Baja, on which we have found numerous porcelain shards and pieces of beeswax (a valuable cargo) but no actual hulls. In the intervening centuries, the shorelines have shifted substantially, evidently burying the wrecks beneath beaches and dunes.

On Baja, we've been searching for galleon traces near one of the world's great beachcombing sites, Malarrimo Beach, where Graham Mackintosh once stumbled on a bar's worth of washed-up booze. Gray whales flock there to calve in Scammon's Lagoon (once a notorious killing ground), turtles return there from Japan, and the giant scythe of Punta Eugenia juts into the Pacific, catching debris from all along the North Pacific's nine-thousand-mile arc. Eddy currents in the point's lee dump this flotsam onto Malarrimo.

In the 1870s, Charles Brooks identified five junks wrecked along Mexico's coast, three in the vicinity of Malarrimo and two farther south near Acapulco. In 1962, beachcombers unearthed a prize that might have come from one of them: a ceramic jug that Smithsonian researchers determined was manufactured in Germany's Westervald region between 1690 and 1710. Perhaps it came from the galleon *San Francisco,* which left Manila in 1705 and never reached Acapulco.

Francisco Muñoz, a veteran local pilot, scoured the Malarrimo desert for treasure until, in 1979, on an old storm beach three kilometers from the Pacific, he found a twenty-foot dugout canoe. With expert help, he eventually determined that it had been carved two hundred years earlier from a small redwood by members of a coastal Athabascan tribe called the Tolowa, now numbering only a thousand, who live near the Oregon-California border among the world's tallest trees. It wasn't the only such canoe from the Pacific Northwest to reach that far shore. Three more have since been found, two of them by the indefatigable Francisco. Twenty-eight years after the first, while searching the dunes near Guerrero Negro for the lost galleon *San Felipe*, I came across another old dugout that had the distinctive bow and stern of one from the Pacific Northwest, resting in two halves in the dunes.

If these canoes had remained on the damp forested shores where they were carved, they would have turned to compost long ago. Instead the dry air and dunes of Baja preserved them, probably for centuries. How did they drift all that way? The answers lie in the complex dynamics of the American Coastal Pathway. It is actually just part of a greater pathway, longer than the Pacific is wide, that arcs along the coast east and south from the Aleutian Islands to the equator. The midlatitude segment, from Southern California to Southeast Alaska, switches direction in fall and spring, in pace with the coastal wind. In winter, what's called the Davidson Current (after George Davidson, who traced its path) pushes flotsam north. Then the wind shifts to blow from the north in summer, and pushes the current and flotsam toward the south.

Some drifters follow the American Coastal Pathway's entire length while others escape it laterally to wander the open ocean. Ordinarily a phenomenon called "small eddy diffusion" insures that drifters released together tend not to spread very much, even when they cross an ocean.

Curt discovers a dugout canoe that drifted from the Pacific Northwest to the vicinity of Malarrimo, Baja California's great collector beach.

Sometimes they do not spread at all; thus we get flocks of plastic ducks at Sitka, an armada of hockey gloves at Vancouver, and pieces of flooring washing up together on Cornwall. On rare, evidently random occasions, however, drifters released side by side can take off in opposite directions; of two drifters released in the Bering Strait, one traveled to Norway, the other to British Columbia.

The spilled hockey gloves of 1994 revealed the seesaw path the dugouts would have taken. After drifting halfway across the North Pacific, the gloves first arrived at Vancouver Island in mid-January 1996. Over the course of the next year, the wintry current pushed them to the north, then the northward currents of summer nudged them south, then the Davidson Current drove them north again. Finally, a year later, some reached Malarrimo; ocean pathways, though slow, are very persistent.

A fall or winter storm probably washed Francisco's dugout down a Mendocino river or swamped it in heavy seas offshore. For the rest of the winter it would have drifted north, like the hockey gloves. Come spring it turned south, reaching Malarrimo by fall. There, a hurricane or tsunami cast it far up on the shore. Other canoes doubtless wait to be found under Malarrimo's sands; alert beachcombers have uncovered entire redwood logs interred there.

The American Coastal Pathway, like other ocean paths, wiggles as it runs. From a hundred miles overhead, satellite photos show plankton patches and slabs of cooler and warmer water within it, meandering from side to side. The pathway's back-and-forth winds eddy so much that objects placed in it are as likely to wander across as to drift downstream, escaping in every direction like trains leaving a switching yard.

But some debris travels much farther along the pathway, and along the wider conveyor of Pacific currents of which it forms a part. Starting out off California or Alaska, Japan or the Philippines, this debris arcs as much as nine thousand miles around the North Pacific before finally getting snagged on Punta Eugenia's hook. Some misses Point Eugenia and the other wash-up beaches and keeps circling, back around to Asia and up the conveyor, looping around the gyre and then round again, beating a slow but steady rhythm.

8. The Great Conveyor

The circles of that sea are laws,
Which publish and which hide the cause.

—Ralph Waldo Emerson, "Celestial Love"

The power of the world always works in circles,
and everything tries to be round.

—John Neihardt, *Black Elk Speaks*

In these days of instant communication, we tend to assume that mysteries are solved quickly. But even in modern times word can travel slowly. More than twenty years after glass fishing floats were introduced in the Atlantic and, later, the Pacific, many who found them had no idea what they were; they could not imagine fishing with glass. Puzzling out the origin of a UFO, an unidentified floating object, is like solving a cold case: a painstaking, often hit-and-miss process. This is the raison d'être of the *Beachcombers' Alert*; it brings the entire beachcombing community in to consult on each case.

Even then, tracing UFOs may take years, and most will probably never be conclusively identified. When they are solved, however, the answers can seem embarrassingly obvious. Beachcombers scratched their heads for years over the unmarked, marble-sized plastic spheres that regularly wash up on shores worldwide. Then, one evening in 1997, I returned to my motel after a day manning the *Alert* flotsam booth at a Sea-Bean Symposium. I happened to look at my toiletry kit and realized what they were: the roller balls from Ban and other deodorant sticks. Beaners now call them "Ban beans."

Many symposium attendees also showed me buoyant round, stonelike beads the color of red brick, about a half-inch in diameter. John Dennis was sure they were of volcanic origin, which seemed a reasonable possibility to me. Then Dutch researcher Gerhard Cadée took up the mystery. In the October 2002 issue of the *Alert*, he reported that he had been finding the beads since 1982 on Texel Island, where they were quite common. He added that pumice otherwise did not occur along the Dutch coast, making a volcanic origin unlikely. But he noted that the beads somewhat resembled certain baked, expanded clay pellets that were produced in Belgium and widely used in potting soil, insulation, and concrete aggregate. And this suggested they were man-made.

I published Gerhard's letter and forwarded it to another newsletter, the *Drifting Seed*. Murray R. Gregory, a specialist in marine litter at the University of Auckland in New Zealand, read it there and, in 2005, wrote to Gerhard that his mysterious grains were indeed manufactured clay pellets.

Sometimes solving one mystery opens another. In 1997 the *Alert* reported on a "horned sphere" that Vern Krause, the beachcomber who found the first washed-up duck in Washington, discovered while gathering firewood. It resembled a miniature World War II submarine mine but was expertly welded of high-grade stainless steel, a costly material that would not be wasted on a mine. It bore no markings. I faxed a photo to the Seattle Police Department's explosives unit, which faxed it to the explosive-ordnance disposal unit at the navy's nuclear sub base at Bangor, Washington. The next day the navy experts removed the sphere from the beach. They took it apart and determined that it was neither explosive nor typical military ordnance, and its horns were actually mounting brackets.

One *Alert* reader thought it resembled a "pig," a scraping device used to clean out pipelines. Another guessed it was a heat-activated fire extinguisher, known by professionals as a "fire bottle," that had been mounted near an aircraft engine. Five years later, after passing from one expert to another, the mysterious sphere finally reached the firm that had made it during World War II, which confirmed that it was indeed a fire bottle. Apparently it had drifted across the North Pacific after coming off a bomber that went down off Japan. But a year or so later another fire bottle washed

Enigmatic ceramic urns that washed ashore in 1961 along the Washington coast, apparently spilled in a supertyphoon that struck Japan in 1959.

ashore about a mile from the first, raising a question that researchers are still trying to answer: Is a downed bomber sitting in the waters off the Washington coast?

Of all the strange things that float across the Pacific, one group of UFOs haunts me to this day, for all my best efforts to track it down: enigmatic ceramic urns that began appearing on the Northwest coast in 1961. I first learned of them in the early 1990s when I read Amos Wood's pioneering 1967 book *Beachcombing for Glass Fishing Floats,* the volume that inspired many of the beachcombers who make my research possible. Wood was by then deceased, but I called his son Dick, who still lived in his father's old house near Seattle. Dick graciously lent me two boxes of material his father had collected for a book he never got to write before cancer took him. Inside, I found notes on about fifty washed-up urns; I have since uncovered another thirty. The facts on them were scanty. They stood one to two feet tall and weighed between twenty and fifty pounds. Their watertight cement caps kept them buoyant. They first showed up on the Washington coast in the spring of 1961 and continued washing ashore, with diminishing frequency, over the next forty years.

As I organized the recoveries by date, I realized I was on to something. It would prove to be the longest-running data set available showing the orbits of a gyre. Trouble was, I did not know where the urns came from.

I asked Jim Ingraham to run OSCURS to see where the urns could have started to reach Washington in spring 1961. Repeated computer simulations for differing times suggested the urns could have drifted from Japan beginning in fall 1959. Sure enough, they had washed up in the usual dispersion pattern of Japanese drifters: mostly along the Oregon and Washington coasts with scattered recoveries in northern California, Hawaii, British Columbia, and Alaska, as far north as the Bering Sea.

I recalled that a First Nations woman in British Columbia once told Amos Wood that natural disasters in Japan often meant good beachcombing years later on Vancouver Island. So I checked the records for big storms in Japan in 1959. Sure enough, Super Typhoon Vera, the most powerful typhoon in Japan's recorded history, hit its southern coast in September 1959, bringing 160-mile winds and causing more than five thousand deaths. I hypothesized that the urns spilled from a coastal vessel caught in the storm. According to my rule of thumb—that 1 percent of Japanese drifters show up in North America—since eighty urns had washed up, the vessel must have been carrying thousands.

Unfortunately, I have never been able to confirm this hypothesis. When I give talks on flotsam, I often show photos of the urns in hopes someone in the audience will report another one. This paid off in Kodiak, Alaska, in August 2007, when I spoke at the Alutiiq Museum, which showcases Aleut culture. A gentleman living in a senior home a block away showed me an urn he'd found in nearby Shelikov Strait in the 1980s and kept all these years. I suggested that he donate it to the museum and added it to my data set.

Amos Wood came to be a student of the ocean through spontaneous curiosity and diligent attention rather than academic training. He grew up in the flatlands of Ohio but, like many machinists and engineers of his generation, came out to Seattle to work for the Boeing Airplane Company in the boom years before and during the Second World War. "He became interested in beachcombing through my mom," his son Dick recalls. "She inherited a beach cabin on the southeast side of Whidbey Island [north of Seattle]. They would walk the beach together in the forties and fifties when you still could find really cool stuff. My dad was always on the lookout for something to salvage and make later use of. This was no doubt due to growing up during the Depression. He brought back quite a bit of dimension lumber, which went into restoring the cabin. But he also had an eye out for the unusual."

From this unassuming start, Wood achieved an oceanographic breakthrough. He recaptured the insights of the great nineteenth-century oceanographers that the twentieth century had forgotten. Back then, pioneers such as Admiral Alexander Becher, tallying message bottles in the North Atlantic, and George Davidson, who mapped the drifts of Japanese junks around the North Pacific, discovered that drifters could circle all the way around oceanic gyres. In 1822, Major James Rennell published maps of current measurements showing the two Atlantic subtropical gyres; Becher's maps of drift-bottle releases and recoveries further delineated the shape and orbital period of the Columbus Gyre. Davidson first sketched the Turtle Gyre in the 1870s. A number of drift bottles revealed the existence of the Antarctic Circumpolar Gyre in the late 1800s.

But with the rise of physical oceanography in the twentieth century, science stopped paying attention to the gyres. Again and again I've noticed that scientists tend to delve into all the detail their tools allow—and that precise instruments and sophisticated technology can actually blind them to the big picture. Message bottles, a relatively crude instrument, enabled the pioneers to postulate gyres. By tracking drifting ships, Naval Observatory superintendent Matthew Fontaine Maury could discern localized currents such as the Gulf Stream and the equatorial current along the edges of gyres. Then World War II revolutionized oceanography, spurring the development of current meters and electronic temperature and salinity measurements. Since then oceanographers have studied ever-smaller phenomena, down to the tiny eddies that produce small-scale friction in the sea, but they have lost sight of the ocean around them. The old insights have languished. The standard postwar oceanography textbook, *The Oceans* by Harald Sverdrup, Martin Johnson, and Richard Fleming, did not even mention the gyres that were the starting point for the science of oceanography.

"Gyre" comes from the Greek word "gyro," meaning "to turn"—hence the familiar sandwich made from meat turned on a spit. In water, a gyre may be a vortex, eddy, ring, swirl, whorl, or maelstrom, any formation around which a drifter can complete a circuit. Each gyre has two parts: a narrow belt of faster water and an interior of weaker, less organized, more chaotic currents. You can see gyres in your bathtub or toilet, in soup swirling in a bowl, in the winds circling the eye of a hurricane, and in satellite photos of the rings that spin off the great ocean currents. One especially volatile current, the Kuroshio, convulses like a loose fire hose, spitting out eddies.

Beltway currents such as the Kuroshio, the Gulf Stream, and the Agulhas ("Needles") off South Africa link up like relay racers to form much greater rings: oceanic gyres. The gyres form continuous loops, like a snake biting its tail, but they are composed of distinct currents, like the vertebrae in that snake's backbone. The Kuroshio joins the Kuroshio Extension joins the North Pacific Current joins the California Current joins the North Equatorial Current joins the Kuroshio to form the vast Turtle Gyre.

After the oceans and continents, the gyres are earth's greatest features. Eleven, each the size of a continent or giant island, cover 40 percent of the world ocean. And yet, until now, no one has attempted to systematically examine the largest gyres as a group. And no one has given them names, other than clumsy locational descriptors such as North Pacific Subarctic Gyre. Smaller eddies that are closely studied have been named. That the gyres have not underscores the lack of attention they have received.

To correct that oversight, and to make it easier to remember the great oceanic gyres, we propose these names, honoring explorers and seafarers—both human and nonhuman, drifters especially—who have circled and traversed their vast expanses. For the North Atlantic Subtropical Gyre we have already proposed the Columbus Gyre, after the first mariner to exploit its currents both coming and going—indeed, the first known to have circumnavigated any gyre. For the North Atlantic Subarctic Gyre to the north, we've coined the Viking Gyre, after the equally precocious mariners who followed it from Europe to Iceland to Greenland to Newfoundland and perhaps beyond. For the South Atlantic Subtropical Gyre, the Navigator Gyre, after Portugal's Prince Henry the Navigator, who founded Europe's first navigation school and launched the Age of Discovery.

The North Pacific Subarctic Gyre becomes the Aleut Gyre, after the fearless hunters who paddled their baidarkas across some of the world's stormiest seas in pursuit of seals, otters, and whales. The much larger North Pacific Subtropical Gyre becomes the Turtle Gyre, after another class of paddlers who cross the widest ocean leaving and returning to their ancient breeding beaches in Japan. And the South Pacific Subtropical Gyre, the Heyerdahl Gyre, after the first fearless explorer-scientist to prove an ancient voyage was possible by reenacting it. The Indian Ocean's gyre is now the Majid Gyre, after the great fifteenth-century Arab mariner and author Ahmad Bin Majid, whose maps guided the Portuguese in their globe-spanning voyages.

We call the Antarctic Circumpolar Gyre, a globe spanner in its own right, the Penguin Gyre, and the large transarctic gyre the Polar Bear Gyre.

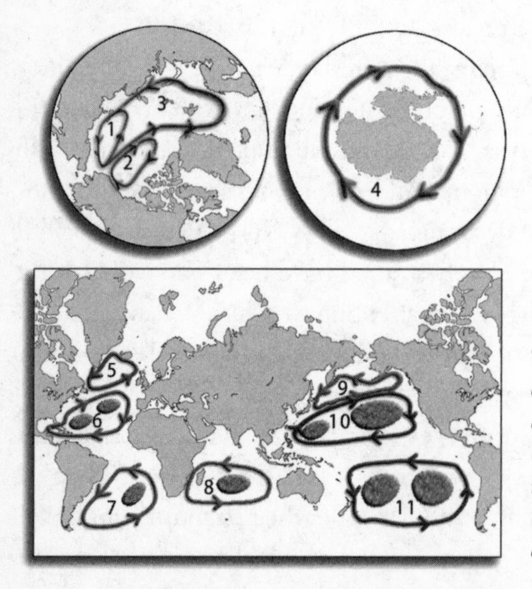

The eleven major oceanic gyres, containing eight garbage patches, are (1) Melville, (2) Storkerson, (3) Polar Bear, (4) Penguin, (5) Viking, (6) Columbus, (7) Navigator, (8) Majid, (9) Aleut, (10) Turtle, and (11) Heyerdahl.

We've named the smaller polar gyre above North America the Storkerson Gyre, after Storker T. Storkerson, the Norwegian explorer who rode an ice island around it. And its Siberian counterpart, the Melville Gyre, not after the author of *Moby-Dick*, though that might suit a more southerly gyre, but after the naval officer George Melville who, as explorer, shipwreck survivor, and oceanographer, did more than anyone to first reveal the movements of it and the Polar Bear Gyre. (For details of the gyres' circumferences and orbital periods, see appendix C, page 235.)

Amos Wood broke the silence about the gyres when he wrote about glass floats circling the Turtle Gyre. And he went further: He pioneered the idea of what I call gyre *memory*. Wood calculated the gyre's attrition rate—how many of the floats it carried were lost (that is, washed up) over a given period of time. He calculated that about 20 percent of a given drift sample would be lost with each orbit. With much more data, I now estimate that, on average across all the world's gyres (except, for now, the frozen Arctic gyres), about 50 percent of a drift sample is lost with each revolution. Like radioactive isotopes, orbiting objects have a half-life, which happens to correspond to one revolution of a gyre.

In the end, the urns proved to be one of thirteen long drift-data sets—along with tub toys, lobster-pot tags, glass balls, submarine mines, sneakers, Becher's and Prince Albert I's message bottles, and NOAA drift cards—whose half-lives I have been able to calculate. Together they show an average dispersal rate of one half-life per orbit. Knowing this, and knowing how long one revolution of a given gyre takes, we can calculate when the last of any group of drifters will wash ashore. Amos Wood calculated that the Pacific's last glass fishing floats will wash up in 2145. I calculate 2177. Neither of us will be around to check the results.

I have, however, been able to tally the intervals between peaks in recoveries of the mysterious ceramic urns. In Alaska I discovered a number of three-year intervals, matching the subarctic Aleut Gyre that sweeps past Kodiak Island. To the south, I found three intervals of five to seven years—the orbital period of the larger Turtle Gyre, which passes from Vancouver

Island down to Baja California. The number of recoveries tapered off in time. Once again, the half-life matched one revolution; about half the urns dropped out and washed ashore with each orbit.

Many observers assumed that these cryptic drifters were crematory urns. But several that were opened contained food of some sort mixed with seawater, not ashes. I still do not know where they originated. But that does not stop them from revealing the cycles of the gyres.

As far back as the 1840s, Alexander Becher had unwittingly chronicled the passage of drifters around a gyre and even compiled data suggesting what that gyre's orbital period was. Of the 152 North Atlantic message bottles he tallied, 80 percent were recovered within a year of their release, on the same side of the ocean, and 97 percent within three years. But six stragglers were only reported four or more years later.

It would be easy to write these six off as having simply lain for years on the beach—except that the periods of time that they "lay" fell into two curious clusters. Four were recovered after nine to ten years (approximately three times three). And two were recovered after fourteen to fifteen years (five times three). None were recovered in the intervening periods. This suggests that these six stragglers were actually marathon runners that had completed, respectively, three and five orbits of the Columbus Gyre—and that one revolution around this gyre takes about three years.

More than a century later, Amos Wood surmised that glass balls circled the Pacific's much vaster Turtle Gyre. In the decades that followed, I located several drifters that had indeed traveled completely around a gyre and wound up where they started. Two of these cases—one an accidental drifter and the other a deliberate one—were particularly striking.

In 1999, I received a letter from a surfer in Haleiwa, Hawaii, named Randy Rarick. It began in typical fashion: "Saw a story in the local paper about glass balls and effects of currents. Mentioned, if one had a story, to write you. So, here's mine." Rarick went on to recall Buzzy Trent, "one of the 'big guns' of the small band of fearless surfers who rode the biggest waves." Rarick built Trent's boards, including one of the last boards he rode. "Being the eccentric individual that he was, he came to me for a special request. It was to be a sleek narrow rocket of a board with the distinctive color scheme of battleship gray with no adornment other than the cigar band from an exotic Cuban cigar that he smoked at the time." Rarick built it and, as usual, signed his name.

In the winter of 1971–72, Buzzy lost the board in a "tremendous wipeout," and a rip current carried it out to sea. In those days before jet skis and rescue helicopters, the only way to retrieve a lost board was to paddle out on another one. "Since this was one of Buzzy's favorites, that is exactly what he did." Hours later, he paddled back empty-handed. Bye-bye, Buzzy's board.

Nearly six years later, in summer 1977, a friend called to tell Rarick an old board—scorched by the sun, bearded with sea grass, and decorated with an odd cigar label—had just washed up on Kauai's north shore, still bearing Rarick's signature. Had it circumnavigated the Pacific or merely meandered within the Hawaiian archipelago for the past five-and-a-half years? Considering how large, visible, and potentially valuable a surfboard is, it seems unlikely one would drift around the surfing islands so long without being retrieved.

Eighteen years later and three thousand nautical miles up-current, another memorable drifter suggested an answer to that question. On December 14, 1995, hundred-mile-an-hour winds slammed the Washington coast. The next day, amid the new debris strewn along oyster-rich Willapa Bay, Brian F. Regimbal discovered a New York Seltzer bottle with its cap so tightly affixed he needed pliers to remove it.

Inside was a note dated February 20, 1990. It opened with a *Garfield* cat face and closed with a smile face. "Hi, My name is Michelle Stone. I'm 10 years old and I'm in the fourth grade." Michelle and her classmates' bottles had been mailed from their school in Eastern Washington to Ilwaco, near Willapa Bay's southern tip, loaded on a fishing boat, and dropped overboard. Michelle recounted how her dad grew wheat and barley, her mom taught preschool, her sister played basketball, and she liked to swim, fish, and ride horses with her cousin Sara. "P.S. write back soon," she concluded.

Six years is a long time for a youngster to wait. Buzzy's surfboard and Michelle's bottle had traveled about the same length of time before beaching near their starting points. And they were just two of forty floaters I've located that appear to have circumnavigated the Turtle Gyre in about six years. Finally, OSCURS clinched the question by reconstructing the odysseys of Buzzy's board and Michelle's bottle according to the winds and waves prevailing at the time. "Round and round they go," noted Jim Ingraham. "Where they stop only OSCURS knows." The board and the bottle followed remarkably similar paths; they even threaded their way between the Japanese islands of Honshu and Kyushu without beaching.

Despite such examples, as of 2007 no one had proven to scientists' satisfaction that flotsam can orbit a gyre. Nor had anyone resolved a basic question: Does flotsam drift around a gyre at the same speed as the water in it? Now, however, OSCURS, satellite-tracked buoys, and our beachcomber network provided the capability to attack those questions. Over the course of fifteen years, from 1992 to 2007, through both my own beachcombing and others', I had gathered seven data sets that yielded orbital periods for the Aleut Gyre. As I collected them, I had no idea that this data would prove critical to unraveling the world's gyres; I just followed my fascination with all things afloat. When I laid out all the data on a spaghetti map—oceanographic jargon for a mass of overlaid lines tracking many different drifters—I found that they delineated the Aleut Gyre's three-year orbit. The pieces in this puzzle included:

Toys. First, of course, came the First Years beavers, turtles, frogs, and ducks. Through thorough beachcombing and scrupulous record keeping, Dean and Tyler Orbison had established a thirteen-year time line for 111 tub toys that washed up near Sitka. As recounted in chapter 4, they found that wash-ups peaked in 1992, 1994, 1999, 2002, and 2004—about every three years.

Sandals. Thank the helpful folks at Nike for yet another data point. They confirmed to me that on January 22, 2000, a cargo box containing 10,224 Nike Baby Sunray II sandals went overboard west of the famous toy spill. All bore a unique code indicating the Chinese factory where they were made and the date they were to be delivered. Six years later I visited Laszlo Hanko, who operated the electric generators for Kodiak, Alaska, and, in his free time, beachcombed. In between jokes about how we could plunge the town into darkness, Laszlo showed me ten sandals he'd found on Kodiak Island in 2001 and 2004—one year and four years after the spill. OSCURS showed that sure enough, the sandals would have first reached Kodiak after ten months, then turned west in the Alaska Current, looped around the gyre, and passed there again.

Bottles. Brian Gisborne of Victoria, British Columbia, maintained an unusual triple career. He operated a water taxi out of Victoria, took tourists out to watch whales and comb remote beaches, and researched marine topics for various government agencies. During this work he came across an old document that I recognized as the official report on Canadian activity during the 1956 International Geophysical Year. At sites in the Gulf of Alaska and along the southern boundary of the Aleut Gyre, oceanographers released 19,449 messages in brown beer bottles. The report presented to the public listed recoveries through 1962, but what Brian had was an office copy updated to include ninety-seven additional bottles recovered through 1972. It showed recoveries peaking every three years along the Alaskan coast.

Pumice. For years Brian mentioned long-ago reports of seaborne pumice, but he could not find the source. Finally he located a 1914 newspaper clipping from the Queen Charlotte Islands describing odd-colored pumice that covered local beaches. I guessed this was ash from the June 6, 1912, eruption of Mount Katmai on the Alaska Peninsula, which poured pumice into Shelikov Strait. Ash scattered on the seafloor confirmed that Katmai pumice had drifted southwest through the strait, toward Asia, and back around the gyre to the Queen Charlottes.

All these drifters circled the Aleut Gyre along an orbit of approximately three years. But was that in fact the gyre's orbital period? Did water travel around the gyre at the same speed as flotsam? I had worked hard to find drifters I thought would mimic water, and I believed that the winds had a negligible direct effect on them. The sandals floated upside down with their soles scarcely exposed, the heavy Canadian beer bottles drifted largely submerged on their sides, and waterlogged pumice barely floats. As for the toys, I discarded their first orbit around the gyre, when they rode high in the wind, and considered only later orbits, after they had become cracked and punctured and barely floated.

Still, to be certain, I needed to measure the movement of water itself—and that's like tracking ghosts. It seemed a water chaser's ultimate challenge.

Then I realized I did not have to chase a water slab in Lagrangian fashion, like a cop pursuing a motorist, as I had on Dabob Bay. If I could obtain a series of temperature and salinity readings taken at a fixed point on a gyre over a long enough period of time, I could find repeating patterns and thus discern the gyre's orbital period—a classic Eulerian approach, like a cop standing by the road monitoring traffic. Then I could test my hypothesis: that flotsam paced water as they rounded a gyre.

In most gyres there are no open-water sites where temperature and salinity have been systematically sampled for several decades. The Aleut and Columbus gyres are the exceptions. In the Columbus Gyre, these qualities have been measured monthly since 1954 in the same area near Bermuda where I chased meddies in the 1970s during POLYMODE. And in the Aleut Gyre, a solution appeared in 2005 at Sitka, during the Pacific Pathways IV conference—over beers.

Sitka is a schizoid village that makes its living off two very different kinds of vessels, fishing boats and cruise ships. Oceanographers prefer the fishermen's hangouts, such as Ernie's Bar. It was there I hoisted a couple pints with my longtime colleague Tom Royer, a trailblazing oceanographer at the University of Alaska. Several times each year for thirty-five years, Tom had measured temperature and salinity in the top hundred meters of water off Resurrection Bay on the rugged Kenai Peninsula. It took heroic efforts to maintain this time line in such rough waters, especially when the same storms that knocked containers off cargo ships howled in from the Gulf of Alaska. Now Tom shared what he'd compiled, and I described the Orbison time line of tub-toy wash-ups. In a flash, water chasing and flotsam tracking fused in my mind. Why not see whether the timing of the toys washing up near Sitka matched that of the slabs passing by Resurrection Bay?

Tom's data and OSCURS's capabilities provided a much easier way to track water slabs than pursuing them with sensors dangling off a boat. Jim programmed OSCURS to mimic chunks of water drifting around the Aleut Gyre and passing Resurrection Bay. Then, using Tom's data, a mathematical technique called spectral analysis calculated what orbital periods could have yielded those patterns. Lo and behold, these periods corresponded to the three-year spells observed between wash-ups of the bath toys, sandals, and beer bottles; the match was so close as to provide a 95 percent assurance that it was not random.

I now appreciated how much washed-up flotsam could tell us about the vast watery world around us; it could reveal the rhythms of an oceanic gyre. With this confirmation, I felt confident that flotsam cycles could likewise reveal the rhythms of the other gyres where we did not have a long series of temperature and salinity readings for comparison. Through them, we could hear the heartbeat of the floating world.

To be sure I'd not gone crazy, I submitted my findings to a peer-reviewed journal, the gold standard in science—in this case the leading oceanography journal, *EOS*, named after the Greek goddess of the dawn. My article, entitled "Tub Toys Orbit the Pacific Subarctic Gyre," opened provocatively, noting that scientists today did not believe what oceanographic pioneers had deduced more than a century ago: that flotsam could orbit an entire gyre. And it proceeded to show how the toys had done just that.

Offbeat though the article was, *EOS*'s reviewers and editors raised no issue with its science. This was the first time, out of dozens of papers I've published in such journals over the course of four decades, that I did not have to answer any criticism. "It is truly refreshing to see rich ideas derive from simple (but painstaking) analysis," one anonymous reviewer wrote.

"Because of its title (comical), content and profound conclusions, I read this paper several times, looking for serious flaws, but each time I enjoyed it more and gained more confidence in the authors' conclusions."

The Aleut Gyre was however just one of eleven gyres in the floating world. I knew far less about the other ten. If I had ten lifetimes, I might reach this degree of proof for all of them. Instead I searched for a simpler way to estimate a gyre's orbital period. I thought back to a mantra of elementary physics I'd learned in eighth grade. "Distance equals rate times time." Translated to present circumstances, the orbital period of a gyre equals the distance a floater drifts around its edge divided by that floater's drift speed.

I tallied all the long-haul drifters along the Aleut Gyre—twenty-eight, most of them not included in the *EOS* paper—about which we knew exactly how long and how far they had drifted. From that data I calculated their flotsam speeds, then divided those speeds into the gyre's circumference. The average result reached by this third method of calculating, with nearly the same degree of statistical certainty, was that the Aleut Gyre's orbital period was 2.97 years.

Sixteen years after my mother first called my attention to the stranded Nikes, I had finally found a method by which I could use flotsam to calculate the orbit of each gyre in the floating world. I know that if she had lived to see what her curiosity set in motion, she would be happy.

Together, the eleven gyres cover as much area as all the land on earth. On average they are as large as the United States, though some are much larger than others. They tend to be oblong—three to six times as wide as they are high. All told, it would take seventy-one years for a drifter to circumnavigate all eleven, a total distance of 91,050 miles. Set aside for the moment the Arctic Ocean, where icepacks slow the movement of water and flotsam by 90 percent and give its three gyres much longer orbital periods. The total orbital periods of the gyres in the other three oceans are proportional to the surface areas of those oceans. The area of the Pacific is double that of the Atlantic and a little more than double that of the Indian Ocean. The total periods of the Pacific gyres are almost exactly double those of the single Indian Ocean gyre and the Atlantic gyres.

The intercontinental gyres' boundaries are defined by land masses along some stretches, hence rigid. But for much longer distances they are flexible, defined by currents that can shift and pulsate like the walls of living cells. As they turn, they throw off smaller whirls—satellite gyres. These transitory eddies wander in and around the fixed gyres, and flotsam can detour along with them.

Satellite eddies aren't the only elements complicating the gyre outlines. Three years after the great toy spill, as we tracked the tub toys around the Aleut and Turtle gyres and Jim updated his OSCURS animation of their movements, we realized that we were looking at two kinds of orbits. Some toys drifted around the gyres' peripheries while others made shorter loops inside it.

As these animations showed, the gyres are not simple circles but rather wheels within wheels, orbits within orbits. Some, such as the suborbit around the Azores Islands, collect flotsam; for these I coined the term "garbage patches," and we'll consider them in all their ghastly glory later. OSCURS showed that the tub toys traveled various orbits and suborbits around the Aleut Gyre: two outer orbits, 7,300 miles and 6,800 miles in circumference, and two suborbits of 4,000 and 5,500 miles. All these orbits follow the same route at their eastern end, along the Alaska coast, but then loop progressively farther west. The shortest reaches only to the middle Aleutians, while the two longest stretch all the way to Kamchatka.

No matter which orbit they fell into, however, the toys drifted at nearly the same speed, 5.1 to 5.6 miles per day, so their orbital periods varied largely according to distance. They took 2.2 years to complete the shortest suborbit and 3.5 years to complete the longest outer orbit—an average of three years, with a lot of noisy but insignificant variation.

What's remarkable about these suborbits is how close they lie to each other—close enough to affect each other's movements. In high school I studied airplane motors and learned that pilots had to adjust twin propellers to exactly the same speed; a difference of a few revolutions per minute would produce a low-frequency *wow-wow-wow* that could shake a plane

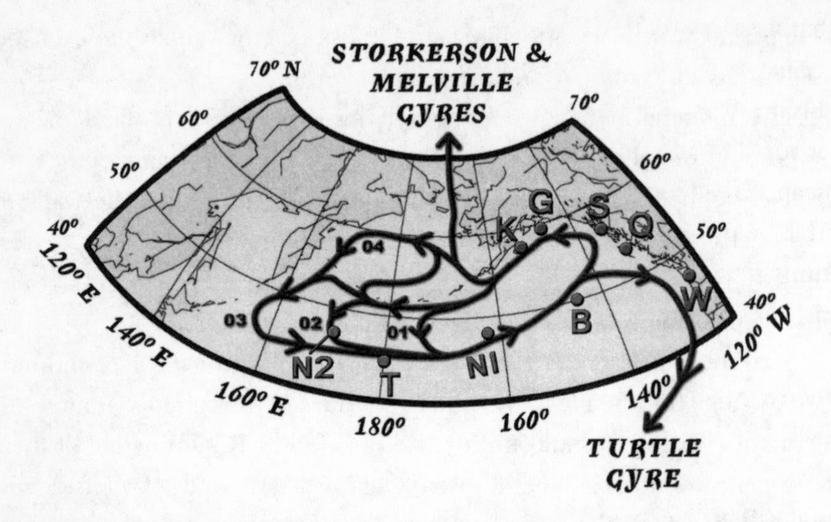

The four orbits of the Aleut Gyre. The outermost and most definitive, O4, has an orbital period of about three years. Open-water dots mark drifter release sites: B, beer bottles at Ocean Weather Station P; K, pumice from Mount Katmai; N1, spilled Nike sneakers; N2, spilled Nike children's sandals; and T, spilled tub toys. Shoreside dots indicate where drifters have been observed: G, water slabs at Resurrection Bay; Q, toys in the Queen Charlotte Islands; S, toys at Sitka, Alaska; and W, toys in Washington.

apart. One reviewer of my EOS paper visualized these suborbits as wobbling like old vinyl records left out in the sun, and when two orbits' wobbles passed each other they would bump together, generating a "beat."

Since all orbits circle at the same speed, it can take decades for one to overtake another by a full revolution and produce a beat. For example, the second and third orbits within the Aleut Gyre, with periods of 2.8 and 3.3 years respectively, "beat" every 18.5 years. This is also the period between lunar nodal tides—the tides that occur when the moon's path crosses the path of the sun in the sky. Many researchers have examined decadal periods in an attempt to understand environmental variability in the North Pacific, such as the famous twenty-to-sixty-year-long Pacific Decadal Oscillation, which produces alternating periods of relatively warm and dry and cool and wet weather in western North America. Some think that the lunar nodal tide is an important factor in causing this oscillation.

Not only are the gyres wheels within wheels, they also function as interlocking gears. The great Penguin Gyre circling Antarctica is a planetary gear, meshing with the southern gyres of the Atlantic, Pacific, and Indian oceans. Where two gyres meet they interlink and hand off a fraction of their flotsam. Most flotsam continues in the gyre it started in because the window for such a handoff is short in both time and space. For example, toys can escape from the Aleut to the Turtle Gyre along only 15 percent of the Aleut's orbit and during just 15 percent of the year, in July and August, when the summer winds blow to the south, nudging flotsam out of the gyre's southeast corner. If these windows don't coincide, flotsam will stay in its orbit.

To visualize this hand-off process, consider one of the most epic (though unconfirmed) voyages in flotsamological lore: the bottle that the eccentric sewing-machine heiress Daisy Alexander reportedly tossed into the Thames River in 1937, containing a will granting the finder half her $12 million estate, which a bankrupt restaurateur found on San Francisco Bay twelve years later. From the river's mouth, the bottle would have turned right, making a J-shaped counterclockwise loop around the North Sea, as confirmed by thousands of bottles released to study fisheries since 1897. North Sea water has been traced far into the Arctic Ocean by tracking two radioisotopes—iodine-129 and cesium-137—discharged from the nuclear fuel–reprocessing plants at Sellafield, England, and La Hague, France.

The Arctic Ocean is a sea of drifting hard water, stirred by the winds like ice cubes stirred in a glass. Though it contains only 1.5 percent of the global ocean's volume and 5 percent of its surface area, it receives 10 percent of the world's river runoff. Most of this fresh water flows from Siberia, producing vast quantities of pack ice. The main flow of this ice, the Transpolar Drift Stream, runs swiftly (for ice—a mile or two a day) over the North Pole and down into the Atlantic between Greenland and Europe. As it proceeds, the stream pushes aside warmer waters rising from the Bering Sea and the Atlantic. This produces two countercurrents: the Storkerson Gyre, which circles above Alaska and Canada, and the Melville Gyre above Siberia.

After exiting the North Sea, Daisy's bottle would have ridden the Melville Gyre eastward to the Bering Strait—a journey of six to eight years. From the Bering Strait, it could have continued along the Transpolar Drift Stream and perhaps eventually returned to England, or it could have circled the Storkerson Gyre; drift cards released at the same time and place

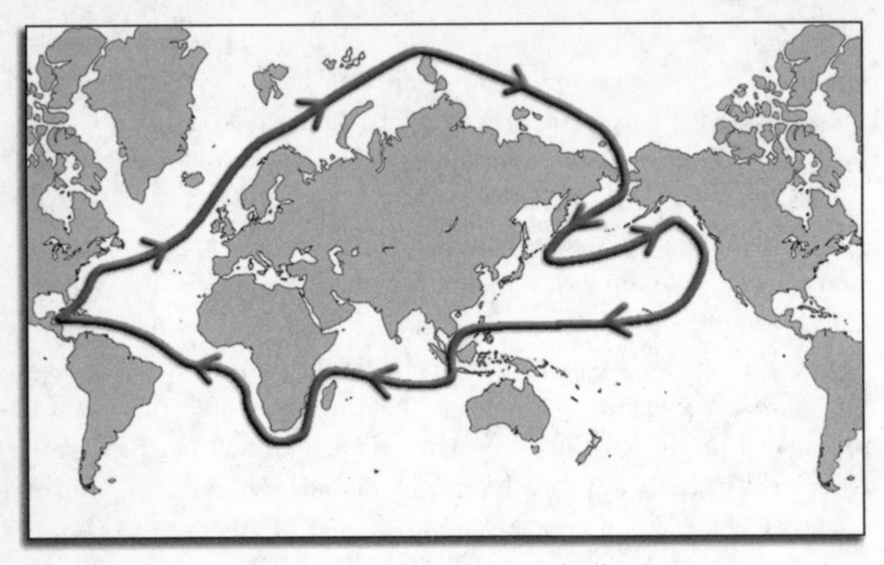

The Grand Tour. Flotsam slings around the world, along the planetary gyres.

have traveled both these routes. Instead, it seems, the windows of opportunity lined up and the $6 million bottle turned south into the Bering Strait and Pacific Ocean, beaching on California.

That route is the northern half of the global conveyor belt, also known as the oceanic Grand Tour, and six confirmed drifters have traveled it. Bottles have also traveled each segment of the belt's southern half, but none are known to have followed the entire route. To complete this journey by proxy is California bottle launcher Alan Schwartz's great dream. "Our ultimate goal is a simple one," he explains, "to cross the Pacific Ocean, thread the narrows of Malaysia and Indonesia, ride surface currents across the Indian Ocean, round Africa, voyage up the Africa coast, cross the Atlantic to the Amazon River, hitch a ride through the Caribbean to the Gulf Stream, travel up the east coast of North America, across the Atlantic . . . and have the message be returned from Ireland or Scotland." Or one of Schwartz's bottles might continue up the Northeast Passage, across the Arctic, down through the Pacific to Japan, and across to North America. It would travel the entire Grand Tour, around the world in twenty years—a human generation.

9. Ashes to Ashes, Life from the Sea

Drift on, drift on, my soul,
towards the most pure,
most dark oblivion.

—D. H. LAWRENCE, "SHIP OF DEATH"

For two decades—one round of the global conveyor belt—Akira Okubo and I worked together whenever we could, on my many trips to New York and his even more frequent visits to Seattle. To ease my access to the facilities at Stony Brook, he arranged to have me appointed an affiliate assistant professor. Together, we published papers on the dispersion of icebergs in the Arctic, mosquito swarms in Iowa, sewage from West Point, and drogues in the Great Lakes. Akira never lost his fascination with the ways clusters of particles scattered and with the mind-boggling possibilities of probability theory. "If you wait long enough, even the most unlikely of events will occur, no matter how small the odds," he liked to say. Once, to make his point, he jotted down in precise calculus a few numbers proving it possible that at some moment between big bangs all the atoms in the universe would reside in just half of it. Even if I could recapitulate his calculations, they would only make the idea more mystifying. Somehow it all made sense when he explained it.

Akira approached everything with the same playful passion. When he stopped by our house we would discuss all manner of things, listen to

Beethoven's symphonies—all nine together on one night—and do jigsaw puzzles. Once he brought along the world's most difficult puzzle, a two-sided mind-twister showing fifty cats on one side and the same image rotated ninety degrees on the other. We worked on it till five o'clock in the morning, when a taxi took him to the airport.

Akira had another great love: Keiko Parker. They had grown up near each other in Tokyo. Both their families were descended from samurai, and hers was wealthy as well. They fell in love when Akira was a teenager and Keiko, ten years younger, still a child. When Keiko wanted to study in America, her parents let her go on condition that Akira act as her guardian.

Both were so stubborn and willful that they knew they could never marry each other. They married others, but remained close all their lives. Keiko translated Akira's book *Diffusion and Ecological Problems: Mathematical Models*—a classic in its field—from English to Japanese, and kept Akira's extensive notes for a second edition. As the years unfolded, she would prove to be his guardian angel.

I meanwhile grew increasingly fascinated with one type of drifting particle, volcanic ash, and its relationship to the sea and, perhaps, life itself. My interest was first sparked by a famous event that I did not even notice at the time, though I happened to pass eerily close by it. In May 1980, Bob Hamilton and I conducted oceanographic observations in northern California's Humboldt Bay. Bob flew back to Seattle while I stayed behind to pack up our field equipment. On May 18, I left Eureka early in the morning and headed up Interstate 5 in a rental van packed with gear; that evening I passed Mount St. Helens. As usual when in the field, I hadn't had time to read a newspaper, and I had no idea that the volcano had erupted. I wondered at the inch of fine ash blanketing the freeway but, exhausted and anxious to get home, drove on. All I could think about was staying awake and not wrecking the van. The next day I read the papers and understood where the dust had come from.

Over the next two weeks, plumes from Mount St. Helens circled the earth; jet pilots changed course to avoid the abrasive particles. Chunks of pumice from the eruption, some larger than a softball, washed down the Columbia River and speedily drifted north as far as Sitka. For years afterward, beachcombers reported pumice washed up along the coast, all the way to Kodiak Island. But it did not appear to round the Turtle Gyre to Hawaii, and I soon figured out why. I scooped up some clasts of pumice at the blast site and set them afloat in a tub of water. After several weeks they became waterlogged and sank, leaving a goop the color of a chocolate milkshake.

Pumice forms when volcanoes eject molten bubbles filled with hot gas, like the bubbles in champagne—except that instead of popping and dissipating, these bubbles harden into volcanic glass, forming airy, porous rocks. Mount St. Helens alerted me to watch out for pumice; since then, I've encountered it on every beach I've combed around the North Pacific. Often novice beachcombers will ask me to identify a washed-up stone they think is unusual. It usually turns out to be pumice—a fact easily verified by tossing it back on the water.

As I studied the floating stones and the volcanoes that produced them, I found that each eruption scattered pumice of a distinct color and particular floatability. The latter quality depends on the size of the rocks ejected by the volcano; smaller clasts absorb water and sink faster than larger ones. St. Helens's small, soft clasts did not drift far. But every eruption also produces larger blocks that can float for thousands of miles, like those that drifted from the Columbia to Kodiak. Pumice from other eruptions has traveled much farther, Krakatoa's crossed the Indian Ocean, and pumice from a 1962 eruption in the South Sandwich Islands circled Antarctica on the Penguin Gyre, then rounded the Heyerdahl and Turtle gyres to reach Hawaii—one of the greatest drifts on record.

Volcanoes have blasted and spewed since the earth's geological infancy, millions of years before there was life. As soon as water gathered, pumice floated on it, crossing the sea through every age and in every climate. Ashes were the original drifters—and what sprawling drifts they formed! On August 26 and 27, 1883, Krakatoa shot three cubic miles of rock over the Sunda Strait. This pumice floated for some twenty-one months and spread across a tenth of the Indian Ocean, traveling as far as Durban in the province of Natal, South Africa; vast patches slowed or stopped the progress of ships. "The main body of the pumice passed Natal in September and October last [1884] and a large quantity was cast ashore there on the 29th of September 1884," one witness wrote. "Yet an immense field is floating about between the Maldives and Ceylon [Sri Lanka], seemingly like the Sargasso Sea is in the Atlantic, only in the one case weeds and in the other pumice, for miles and miles the sea is covered with a thick coat of pumice, on top of which are numerous crabs and beneath are very many strange fish making it a home and shelter."

Pumice and ice together overran southern Iceland's coastal waters in 1362, when the volcano Vatna exploded. "The Knappafel Glacier gave way and flowed down into the sea," *The Annals of the Bishopric of Skáholt* recorded. "And this also happened: Great heaps of pumice were drifting outside the West Fjords so that ships could scarcely make their way through." The floating pumice reached the West Fjords, halfway around the island, riding the same coastal current that bore high seat posts and Old Kveldúlfur's coffin on the Vikings' search for harbors.

Seventy-four thousand years ago a much greater blast created Lake Toba at the opposite end of Sumatra from Krakatoa. It spread a ten-thousand-square-mile sheet of pumice, more than a thousand feet thick in places, across the surrounding land and waters. Today, Toba would be one of more than 370 active volcanoes circumscribing the Pacific in the Ring of Fire. With that tally in mind, I asked myself, how much pumice would it take to cover the Pacific, or even the entire world ocean? This is like gauging how many gallons of paint you need to coat a house; it doesn't take much to cover a large surface with a thin layer. Likewise, waves and winds disperse pumice clasts over a wide area.

Each increment in the standard volcanic explosivity index (VEI) represents a tenfold increase in the volume of pumice and ash ejected in an eruption. Krakatoa, a VEI 6 eruption, covered a tenth of the Indian Ocean, or 2 percent of the world ocean. So it seems likely a VEI 7—four of which have occurred in the last ten thousand years—could cover the entire Indian Ocean. And a Toba-sized VEI 8, the maximum level, might produce enough pumice to cover most of the world ocean.

Imagine that ocean's beginnings. As the earth cooled, widespread volcanism thickened the atmosphere and vast oceans appeared. As soon as they did, currents infested them, spreading whatever they could bear. Meanwhile, in the estimated 900 million years between the formation of the earth and the appearance of the first living organisms, ten thousand VEI 8 eruptions occurred, coating the ocean with pumice. Only pumice and ice washed up on the lifeless shores.

Even today, finding a piece of pumice on the beach makes me imagine what that time must have been like. I'll throw the pumice into the surf and stare as it floats, trying to visualize the sea's earliest days. My mind loses control of my eyes and my left eye wanders outward, throwing my vision out of focus. And my inner voice makes itself heard.

And so an idea came to me as I watched a piece of pumice from the South Sandwich Islands eruption of 1962 float, not in the surf but in a plastic tub atop the tank of my basement toilet. I'd dropped the pumice and a few walnuts, which like the rock were deeply fissured, in together to see how they floated, and for how long. Flushing the toilet jostled them like waves stirring the ocean's froth. After a few weeks a scum formed, coating the stone and nuts. Here, I realized, were the perfect conditions for life to evolve: a slowly absorbent sponge, a cluster of floating test tubes.

The time was June 1993. Tub toys were stranding by the thousands along Alaska. On Tuesday, June 22, Jim and I finally managed to simulate their drift with OSCURS. That day Akira arrived on his forty-ninth visit to Seattle, and we had lunch at the Santa Fe Café, a block from my house. Over taco soup, I mentioned the idea that had come to me, that life might have originated in pumice floating on the primordial sea. By the end of lunch we'd written out a detailed scenario for this evolution. I hurried home and typed out my notes. Three days later we submitted a letter to *Nature*, the leading journal of the life sciences, entitled "Origin of Life in Floating Pumice."

In it we explained how long before life arose and soon after the ocean came to be, drifting rocks appeared—the first flotsam. The massive volcanism of earth's infancy dusted the primordial ocean with vast quantities of pumice, which could then float for years. As a permeable matrix, composed mainly of silicon dioxide and voids, this pumice would adsorb a variety of chemicals. These chemicals would have been exposed to high concentrations of energy via solar radiation and direct lightning strikes, which fueled the formation of the amino acids and proteins that are the building blocks of life. Black pumice would have absorbed heat especially readily, promoting chemical reactions. Such reactions generally proceed more rapidly on surfaces than in free water, and because pumice is fractal—its effective surface area is very large—it would have afforded an especially rich bed for them.

Sedimentary records suggest that biodiversity was quite high in the primeval era. Such diversification requires sheltered or segregated environments in which life forms can evolve along many different paths. And what environment could be more sheltered than the interstices in drifting pumice? Each chunk was a floating island, an isolated world of micro-environments. Billions of rocks allowed much more diversity than the swirling, homogenizing seas.

When pumice became stranded on land, evaporation and the resultant concentration of adsorbed chemicals may have further aided the growth of microbes. If pumice became waterlogged and sank, or if these microbes left the protected interstices, they would compete with others in the turbulent ocean, where the best adapted would overwhelm the less competitive, reducing diversity.

Our model resolved some of the problems that dogged familiar notions of life's emergence in the primordial sea: How would energy from above penetrate the dense water? Why wouldn't precursor amino acids be dispersed before they could assemble into proteins and organisms? If our speculation was correct, we suggested, some corollaries should be explored. Similar processes might be continuing now. Pumice drifting today could be examined for the presence of adsorbed complex chemicals. Cutting cross sections of ancient pumice might reveal archaeobacteria in the interstices—much like ancient insects preserved in amber.

Laboratory experiments had subjected seawater to artificial lightning in the presence of atmospheres matching ancient conditions. We suggested interjecting pumice into these experiments, as well as replicating certain oceanic processes that might also have abetted the emergence of life. In the primordial seas, wind-generated waves doubtless produced foam, which accumulated in windrows. These windrows would have collected drifting pumice, yielding a mixture of froth and pumice—a fruitful medium for forming complex chemicals, especially if lightning were also applied.

And so the top microlayer of the infant sea could have been the amniotic fluid, and pumice the floating test-tube rack, in which inert material began its slow transformation into animate beings that love, dream, think about themselves, and reinvent the universe. We urged readers to undertake the experiments that might prove this possible. But *Nature* rejected our manuscript, the editor noting that it seemed a good idea but was too speculative at this stage and required additional investigation. We agreed; our intent was to provoke just that sort of investigation. Without funding we could not execute the experiments we had suggested.

On Christmas Day, 1994, Akira arrived on his fiftieth visit to Seattle. For several years he had suffered from severe abdominal pains, and we'd all urged him to get medical attention. But he was of the generation that did not trust doctors, and his samurai pride made him doubly resistant to pleading for help. Now when the pain struck he would double over and wait for it to pass.

Soon afterward, back in New York, Keiko finally persuaded Akira to see a doctor, who directed him to get a colonoscopy. It showed that he had advanced colon cancer, the disease that afflicted my father and killed both my grandfather and Susie's mother. Worse yet, the cancer had spread to his liver. The surgeon cut out 85 percent of the liver, the most that could be excised and still allow the organ to regenerate. Afterward, the doctor said he'd removed all the cancer. A few months later it returned, now inoperable, and the surgeon refunded Akira's money. Ever since then I've gotten on my soapbox every chance I get, urging friends to get their colonoscopies as soon as possible.

That summer Akira retired from Stony Brook and I attended a

OCEANOGRAPHY

THE OFFICIAL MAGAZINE OF THE OCEANOGRAPHY SOCIETY

VOL. 12 • NO. 1

Tribute to
Akira Okubo
1925 – 1996

Program for the 1999 TOS
Scientific Meeting:
Extreme and Unexpected
Phenomena in the Ocean

Akira Okubo, honored on the cover of Oceanography.

symposium in honor of the occasion. He wore sandals, a blue-and-white robe, and a sword in observance of his samurai heritage. In his farewell speech, he said he hoped that he and I would finally get a chance to go beachcombing together.

We never got to go beachcombing, or to follow up on his ideas about the possible arrangements of the universe's atoms, or to finish the world's hardest jigsaw puzzle. Keiko cared for Akira in his last months. He died on February 1, 1996, at the age of seventy-one. As he had requested, his body was cremated; the undertakers incinerated it for several hours at 1,600 to 2,000 degrees Fahrenheit, boiling off sixteen quarts of water and leaving four pounds of dense grey ashes. Three months later Keiko brought Akira's ashes to Seattle, the city he loved. I had told them both how ashes ride the currents across the sea and only very slowly settle to the bottom, and they both decided that was where they wanted their remains to travel.

People living in many different times and places have imagined the soul finding liberation on the currents. It's much more common to dispose of ashes rather than whole bodies at sea, but even this can be trickier than it seems. A ship's engineer once told me how he and the rest of the crew had borrowed a tanker for a few hours to deliver the ashes of its recently deceased skipper to the sea. Unfortunately, the captain rejoined his ship rather than the sea; the wind blew his remains about and they stuck to the white superstructure, which the crew had just painted. The crew chipped the ashen paint off and, apologizing to their Old Man, completed the ceremony by scattering the white flakes at sea.

If you burn the boat with the body, you won't have to worry about ashes sticking to the paint. Immolation at sea has a long and hallowed history: Vikings, Northwest Indians, Fijians, Maoris, and the Dyaks of Borneo all cremated their chiefs aboard sepulchral ships set adrift. Today the average person can afford these rites, and a seaworthy urn has replaced the floating pyre: the Velella, a patented, football-sized plastic container that biodegrades as it drifts, releasing the ashes over a two-year period so its occupants can take a posthumous ocean voyage.

There's plenty of room at sea for everyone who might want to take such a voyage. If all the six-and-a-half-billion humans now living were to be incinerated and scattered over the waters, they'd form a layer a thousand times thinner than the finest human hair.

On Friday, May 10, 1996, I had the sad honor of adding the ashes of Akira Okubo, my friend, mentor, and collaborator, to that oceanic nanolayer. Four of us—James J. (Jim) Anderson, Keiko Parker, Michiyo Shima, and I—made the three-hour drive from Seattle to Diablo Lake in the North Cascades, right below Ross Lake, my father's favorite retreat. Jim had worked closely with Akira at the University of Washington's Center for Quantitative Science, and he and I had been in graduate school together. Michiyo, like me, was one of the many students worldwide whom Akira had befriended and inspired.

At Diablo Lake we hiked in past a locked guard gate to the visitors' dock moored a few hundred yards behind 389-foot-tall Diablo Dam, once the tallest dam in the world. It was the middle of the week, and the dock was deserted. I removed the black plastic box containing Akira's ashes from my backpack. We sat cross-legged on the dock and read poetry. I recited "A

Little Lamp Went Out" by Friedrich Ruckert (which Gustav Mahler set to heart-rending music). And we threw Akira's ashes into Diablo's chalky green waters, clouded with fine clay particles borne by the melt water from fourteen shrinking glaciers.

I watched the little cloud that once was Akira drift toward the dam; it was hard to discern in the milky water. I knew that the finest ashes would swirl through the turbines and travel down the Skagit River, out into Puget Sound, and finally into the Pacific. My thoughts went back to the Christmas snow globe my father brought home from one of his many chocolate-selling trips. I remember endlessly shaking it and watching the white flakes settle slowly through the water. Now, as I sat atop Diablo Lake, it seemed to me a miniature ocean full of settling ashes.

I wondered if any of Akira's dust would reach Tokyo, his boyhood home. I knew that some of the ashes were even finer than the glacial clay, which settles through seawater at the rate of just one millimeter per day. It would take forty thousand years for them to sink to the lowest point in the Pacific Ocean, seven miles deep. In my brown daybook I calculated how long it would take them to settle a hundred feet below the sea surface, the depth to which sunlight typically penetrates. (Had the circumstances been reversed I know Akira would have calculated how long my ashes would remain suspended.) The answer, eighty-three years, astounded me.

My view of time changed as I came to understand the rhythms of the gyres. I imagined them as clock faces with hands of flotsam. In the coming decades, billions upon billions of Akira's particles would drift tens of thousand miles around the world ocean, first southwest along the California Current past Hawaii, then westward on the North Equatorial Current to the Philippines. There they would turn right along the mighty Kuroshio and finally reach Tokyo Bay. They would orbit the Turtle Gyre fourteen times before falling below the light of day. Some would escape to drift around the other ten oceanic gyres. They would wash up on every shore around the world. Remember, the next time you walk along a beach: Your bare feet brush the atoms of everyone cremated and buried at sea.

This passage suited Akira far better than a hole in the earth would. As Countee Cullen said in his epitaph for Joseph Conrad,

They lie not easy in a grave
Who once have known the sea.

I felt moved to verse myself:

Earthen graveyards
Distinguish rich from poor.
The sea, the greatest cemetery,
Makes no distinctions.

Ever since then, the expression "from dust to dust" has held a special meaning for me. We were born in the dust of volcanoes; we return to dust and become continental drifters ourselves.

Six years later, *Oceanography* magazine finally published our letter hypothesizing that life originated on floating pumice, in a special issue dedicated to Akira and filled with tributes from his colleagues, students, and friends.

For years, in forensic investigations and my own research, I had traced the drifts of strangers' remains. Now I had reached the stage in life when the ashes of those I loved filled my thoughts. Six months after Akira's passing, on July 25, 1996, my father died. Perhaps the marching orders he issued to his recalcitrant cells each morning had worked; despite Parkinson's, cancer, heart disease, and all the other illnesses that afflicted him, he reached the ripe age of eighty-one. My mother and father had agreed that their ashes should be released together in Ross Lake, so I kept his in my desk drawer through one turn of the Turtle Gyre. Ashes naturally harden into blocks over time, and I worried that I might not be able to scatter Dad's when the time finally came.

As I considered my parents' and friends' wishes for their final remains, I asked myself, Why do so many people want to be scattered on the sea, to live along the shore, to comb beaches and send messages in bottles? I remembered my father barking orders to his cells and wondered if individual cells might indeed have a sort of awareness—a body-wide remembrance of the mother sea, a yearning for the medium in which the first living cells came to be.

A parallel yearning plays out even in creatures that dwell in the sea. Like humans migrating to the shore, they cycle back to their places of origin. Salmon return to spawn in their native streams. Sea turtles lay eggs on the beaches where they hatched. Gray whales return to calve in the lagoons where they were born.

In 2008, I was tracing the drift of a buoy from Ginoza, Japan, to Washington when the inner voice whispered to me about another connection. Years earlier I had tracked several such buoys, known as fish attraction devices or FADs, which anglers set to concentrate fish for catching. Why, I had long wondered, were fish attracted to them and to all manner of other drifting objects? The prosaic explanation is that some feed on the colonies of other organisms that grow on these floaters and others feed on the feeders, on up the food chain. But might fish and other creatures, including us, return to drifting objects because that's where life originated—each of our souls a bottle floating in the amniotic fluid of our marine mother's womb?

Keiko saw her death coming years ahead and planned carefully. She fulfilled her pledge to Akira, seeing the second edition of his landmark *Diffusion and Ecological Problems* published in 2002. Later that year, six years after Akira, she died of a burst carotid artery caused by a hereditary condition.

Keiko's love of Akira lived on in her will. She requested that her ashes also be scattered in Lake Diablo so her remains could follow his on the currents to Tokyo. And she left $2,000 specifying that seven friends—Dr. David Strand, Dr. Bridget Duffy, Dr. Jim Anderson, Donna Kinkade, Fred Helmholz, Susie, and I—should hold a wake over dinner at Campagne, one of Seattle's finest restaurants, in the Pike Place Market. The instructions were clear. We had to spend the whole $2,000.

David and Bridget flew to Seattle with Keiko's ashes. On Monday, August 19, 2002, we planned to scatter them at Diablo Lake, but our plans went awry and we wound up at Campagne. It was a hot August evening, but we lucked out and secured a table on the patio, just off Post Alley. We set out Keiko's remains as the centerpiece in an elegant little white ceramic bowl and launched into our sumptuous feast and our memories of Keiko and Akira. We rolled with laughter for hours, each of us sharing tales about them the rest had never known while they were alive. We finally finished dinner, with $1,000 left to spend. What could we do? We informed the waitress of our predicament, and naturally she was glad to help us spend the last thousand on fine chardonnay (Bugey Altesse Peillot) and champagne (Jacques Selosse Blanc de Blanc Avize, one of the most distinctive of all champagnes). So exquisite were they that Susie jotted down the appellations on a card that she treasures to this day.

Of course we got sloshed, as I'm sure Keiko intended. Finally we got around to her remains, sitting there before us. I described the astoundingly slow settling rates of fine ashes and said just a pinch from the bowl, containing trillions of particles, would do to launch her around the Pacific. Bridget and David planned to travel the next day to the picturesque San Juan Islands and sprinkle a few ashes in their waters.

Taking the situation in hand, Jim Anderson lifted the bowl's cover. To our dismay, it contained not ashes but a vertebra, which clearly would not float. How to satisfy Keiko's desire for her ashes to drift like Akira's back to Japan? Fueled by the wine, Jim took the vertebra and easily crushed it in a glass of water with a knife. A fog of fine Keiko dust now floated there. I took the solution one step further, pointing out that we could satisfy Keiko's wish right then and there. "Remember Akira's first work in Seattle, on the West Point sewage treatment plant?" I asked. The toilets in Campagne flushed to that plant. Within a few hours, Keiko's ashes could be in Puget Sound and on their way to Japan. David leapt up with the water glass and we gathered around a toilet located just inside the restaurant. With heads bowed and hands crossed, we launched Keiko on her journey around the Turtle Gyre.

As it happened, Akira's ashes had begun drifting in May 1996, six years—one average orbit of the Turtle Gyre—earlier. They had now rounded back to Washington just in time to be joined by Keiko's. She and Akira are finally together, and will remain so virtually forever.

A beachcomber named Marty Terwilleger once told me he carried along a pinch of his father's ashes whenever he traveled because Dad, a merchant seaman, wanted to tour the world in death as he had in life. The next year my turn came to contemplate the fate of my own parents' ashes. On October 8, 2003, my mother passed away, and five days later the funeral home called to say her ashes were ready to pick up.

To fulfill my parents' final wishes, I rented three cabins at Ross Lake Resort: one for Jim and Jan White; another for my brother Scott and his wife, Karen Ebbesmeyer; and a third for my cousin Bob Blewett and his wife, Ann, their daughter Lori, and her two kids. Getting to Ross Lake is no picnic. We drove across Diablo Dam, past the visitors' dock—where we had sprinkled Akira's ashes—and on to a tugboat landing. The boat arrived and took us to the base of Ross Dam, where a flatbed truck waited. We offloaded our gear and food (you have to pack your own at Ross Lake) and lifted it onto the flatbed, which carried us over the dam to yet another dock. There the resort's manager picked us up in a catamaran and, after a fifty-mile-an-hour dash, dropped us at our cabins, resting on their ancient cedar pontoons. Once again we were floating.

The next day we set out in a twenty-foot skiff. Scott and I had neglected to bring a container from which to scatter the ashes. We scrounged a white five-gallon bucket, mixed our parents' ashes together, and added water so the wind would not kick them up. As I knew from lore and experience, putting ashes over the side of a boat is tricky. They are abrasive, just like volcanic dust, and can harm an outboard engine if they get sucked into the water intake. We slipped the slurry over the side and our pilot gunned the motor to get out from the little cloud that rose around it.

The last I saw of my mother and father, they were intertwined, a white cloud in the sparkling emerald waves. They'd fallen in love at first sight in high school, and now they would never part. Through tears, I watched as the cloud faded and vanished, a mile or so down the approach to Ross Dam. It would circle the Turtle Gyre with the clouds that were once Akira and Keiko—not to mention the dust of bathtub duckies and other industrial flotsam that, being made from organic hydrocarbons, also derived ultimately from pumice floating on the primeval sea. Ashes at the end of life had taught me much about the beginning of life.

10. Junk Beach and the Garbage Patch

The sea is the conscience of our civilization.

—PHILLIPE COUSTEAU

Will all great Neptune's ocean wash this blood
Clean from my hand?

— WILLIAM SHAKESPEARE, *MACBETH*

All through the 1990s, even as I launched the *Beachcombers' Alert*, developed the Alert Network, tracked sneakers and tub toys and other drifters, and puzzled out the movements of the gyres and the possible role of life-incubating pumice, I continued working three-fifths-time as a consulting oceanographer. Finally, however, I came to a point when I could no longer reconcile the facts my investigations uncovered with the uses my clients wanted to make of them.

Bob Hamilton, my partner and boss at Evans Hamilton, was always good at getting to the heart of things. Around forty years ago, at the start of my professional career, Bob spoke to me over that great elixir of environmental truth, beer. For others, he said, oceanography was a job, but for me it was a calling. It was a simple, obvious point, but looking back now I can see that that realization started me on my way to retirement just as I was setting out.

Most of my oceanographic work—from designing offshore oil platforms to placing outfalls for sewage treatment plants—was commissioned to support engineering projects in the ocean and in Puget Sound. I sought the most scientifically valid answers to the questions I was posed, answers I could defend in good conscience. Often these answers conflicted with the engineers' preconceived designs; they expected the environment to behave in accordance with their notions.

When I began working offshore for Mobil, the oil industry was still willing to listen to data. The seventy-two-foot wave we measured during Hurricane Camille prompted it to begin designing for seventy-five rather than fifty-five feet. Twenty years later, the four-knot currents we measured in Loop Current Eddy Nelson convinced the industry to take more notice of eddies in deepwater operations.

But as the years passed, I encountered growing indifference to environmental concerns. I suggested that we suspend fishing for salmon on the Columbia River to let them recover and fish the far more numerous shad instead—and was drummed out of NOAA's salmon-recovery meetings. During oil-spill studies in the Strait of Juan de Fuca, I raised concerns about oil spreading underwater, and colleagues representing the oil industry told me to get lost. In studies of Puget Sound, I explained how the Sound is a dynamically closed system, with most sewage recirculating within it rather than exiting directly. No one in a position to remedy the situation cared then, and no one does now. These findings are not in dispute; they are merely ignored when facilities are planned, and they have not led to any caps on what may be dumped into the Sound. In studies of the Seattle waterfront, I discovered that propeller wash from state ferries idling at their dock stripped away fine sediments and disrupted estuarine flow. The state Department of Transportation hired a competing oceanographer and directed him to find ways to discredit my findings. But they withstood the assault, and have since been published in peer-reviewed journals.

I also served on two governors' advisory panels. In 1993, Governor Booth Gardner appointed me to the Marine Science Panel, with scientists from both Washington state and British Columbia, convened to assess the marine environment in shared U.S.-Canadian waters and recommend action to protect it. We issued our report the next January at an impressive symposium in Vancouver, and—as near as I could tell—it was neatly filed away and forgotten, like so many others.

Gardner's successor, Mike Lowry, and British Columbia's premier got into a dispute over Victoria's sewage outfall, which dumped raw sewage into the Strait of Juan de Fuca. They settled it by appointing a blue-ribbon panel to address transboundary issues in the strait; each appointed three members. I had written about the outfall, which evidently was what got me appointed. Once again we met and, over the course of a year, wrote a report, which was presented to the public. Washington appointed several working groups to evaluate and act on our recommendations. They did their work and it was duly ignored, just as a previous panel's recommendations had been; the whole effort washed away like sewage from an outfall. Now a third effort has been launched from the Washington side. I hope the third time will prove a charm, but I chose to spend my time writing this book rather than sitting in on the endless committee meetings.

I came away feeling that these panels were just political cover, put up to avoid addressing the real environmental problems. As far as I can tell, my participation did nothing except waste my time. But I kept hammering at environmental problems that were not being addressed. I spent years looking into the effects of dams on the Columbia River, Hood Canal (via the North Fork Skokomish River dam), and Whidbey Basin (via the Skagit River dams, including Ross Dam, which impounds my own beloved Ross Lake). I found serious effects in each case, but no serious concern in any government agency. I pointed out the damming effect of the Hood Canal Bridge on the surface outflow of Hood Canal, but the Department of Transportation wasn't interested. My only success was to get transportation officials to recognize the effects of propeller wake on eelgrass and other organisms that depend on fine-grained inshore sediments. But they resented being forced to take this action and blackballed me from doing any further work.

Then, as I approached the age at which I could retire, King County Metro's Brightwater sewage project came along. I was the lead oceanographer in determining where an outfall should be sited. Try as I might, I could not figure out how the effluent could escape Puget Sound from any of the prospective sites, or how it would affect hypoxic (low-oxygen) conditions in the Whidbey Basin, which are as severe as the notorious hypoxia in the Hood Canal. King County officials would not entertain any examination of these questions. They simply said that our budget would not allow further inquiries and that they would look into these questions themselves—which they did not do.

I grew so frustrated that I retired as soon as possible, at the end of 2002, after I turned fifty-nine and a half and federal law let me withdraw funds from my retirement plan. My frustration had built up over the years, from one affront after another to my sense of what was the right thing to do. Those with power to do something felt no urgency about addressing major problems in the Northwest marine environment: the concentration of contaminants in a surface microlayer, sewage escapement from the Sound, the fate of salmon in the Columbia, the Hood Canal Bridge's blockage of circulation in the canal, the effects of dams on estuaries, the reduction of oxygen levels by wood debris in Tacoma's Hylebos Waterway. . . . The list goes on and on, year after year.

Of course, the picture is hardly prettier when you consider marine systems and environments on a global scale, as I have the freedom to do now. But at times there is a grim beauty to the mechanisms via which our actions affect these global systems. Sometime back in the early 1990s, I coined a term that has lately gained wider currency: "garbage patch." These patches are vast, concentrated repositories of whatever floats on the oceans, distributed through the water's top layers. The lightest objects, blown by the wind, travel fastest; waterlogged trunks, old airplane wings, and other bulky relics drag along behind.

Like winds and waves, garbage patches ultimately derive from the sun. Its rays strike the earth most intensely at the equator, warming the air and making it rise. When this warm air strikes the cold upper atmosphere, it cools and starts to descend but gets pushed aside to the north and south by more warm, rising air. Much of this newly cooled air descends in wide bands called Hadley cells, which lie about thirty degrees north and south of the equator, at the latitude of Chongking, Cairo, New Orleans, and, in the south, Durban and Porto Alegre, Brazil.

The Hadley cells descend with special force on vast oblong patches in the subtropical oceans. Each cell produces a high-pressure convergence at the sea surface—an "anticyclone," the opposite of the low-pressure rings around which cyclones and hurricanes form. This high-pressure cell draws air down and pushes it out to the sides. The earth's rotation nudges this ejected air to the right, causing it to circle around the high-pressure ring. The rotation also makes currents head off at a forty-five-degree angle to the wind and turn back into the high-pressure cell, carrying their flotsam with them. This flotsam gathers and spins around a mound in the water formed by the high-pressure cell. A garbage patch forms, collecting whatever natural flotsam floats around it—driftwood, seaweed, pumice—plus whatever we humans choose to throw into the sea or fail to keep out of it.

Only a few garbage patches have been documented. I have tallied eight of them, four in the Pacific, three in the Atlantic, and one in the Indian Ocean. The Turtle Gyre (like the Heyerdahl and Columbus) contains two, which I've dubbed the Great Eastern and Western Garbage Patches. The former, stretching between California and Hawaii, is the largest of all the patches, about half the size of the United States; together the eight patches cover an area more than twice the size of the fifty states.

The high atmospheric pressures above the patches mean weak winds, so sailors cannot cross them. The most famous of all the patches, in the western Columbus Gyre, thus became notorious as a graveyard of becalmed boats, made all the more daunting by its great mats and mazes of *Sargassum* seaweed and flotsam; hence its name, the Sargasso Sea. The other North Atlantic patch, to the east, surrounds the Azores, where Columbus saw the sea beans, bamboo, and other transatlantic flotsam that lured him to America. In the nineteenth and twentieth centuries, when Prince Albert I (the first person to map a garbage patch) and Dean Bumpus released thousands of message bottles on both sides of the Atlantic, one-fifth of those returned were found on the Azores—a vastly disproportionate number from such a tiny territory. Likewise up to 20 percent of the six hundred–odd message bottles that John Dennis recovered from throughout the North Atlantic.

All this historical and oceanographic evidence suggests the Azores are one of several strategically located sites worldwide where converging currents and a leaking gyre deposit enormous quantities of trash. Hence I expect that the Azores contain several major collector beaches. Such beaches are valuable laboratories for surveying the contents and movements of the seas. Some—Malarrimo and Padre Island—I know well. Others I long to visit, from the Azores and Bermuda in the Atlantic to Ducie Island, a particularly remote atoll in the Pitcairn chain, in the far South Pacific. Like the surfers in the movie *Endless Summer,* traveling the globe seeking the perfect wave, I would round the gyres, searching out the world's trashiest beaches.

I might find trashier beaches, but I could hardly hope to observe them in such eloquent company as the pair of illustrious exiles who strolled Los Angeles's Hyperion Beach one autumn day shortly before the start of World War II: Aldous Huxley and Thomas Mann, two literary giants, enthralled and appalled by the brave new world they had landed in while the old world tore itself apart. "Between the breakers and the highway stretched a broad belt of sand, smooth, gently sloping and (blissful surprise!) void of all life but that of the pelicans and godwits," Huxley afterward wrote. "Miraculously, we were alone. Talking of Shakespeare and the musical glasses, the great man and I strolled ahead. The ladies followed. It was they, more observant than their all too literary spouses, who first remarked the truly astounding phenomenon. 'Wait,' they called, 'wait!' And when they had come up with us, they silently pointed. At our feet, and as far as the eye could reach in all directions, the sand was covered with small whitish objects, like dead caterpillars. Recognition dawned. The dead caterpillars were made of rubber and had once been contraceptives."

True to its titanic name, Hyperion was the site of America's largest sewage outfall. Out of sight, out of mind, its operators assumed—an attitude that, though no longer so boldly spoken, still underlies humankind's dealings with the sea. But though the Hyperion pipe ejected its loads a hundred feet below the surface, they did not stay out of sight.

Huxley broadcast this beachfront discovery in his essay "Tomorrow and

Tomorrow and Tomorrow." Afterward, American sewerage operators began screening out what New Yorkers called "East River whitefish." They knew at what hours to be especially vigilant: after midnight nightcaps, in the morning before work, and at 2:30 PM, following afternoon assignations.

Before that change, many local seals and seabirds likely clogged their guts on love's lost labors. But they were fewer surely than the Alaskan fur seals that have succumbed to much deadlier catches on the fabled Pribilof Islands in the middle of the Bering Sea. Once more than 2 million northern fur seals hauled out on St. Paul Island; now fewer than a quarter that number do. Many menaces may be hurting St. Paul's "sea bears." One certainly is—a flood of marine trash. In the early 1980s, St. Paul's beaches were still fairly clean, and fish nets, ropes, and other human-made debris only occasionally peeked out from the driftwood. Now they pile up by the ton (in 2006 nearly eighty tons were removed from the island) and strangle the seals. Between 1998 and 2005, 795 seals were spotted entrapped in lines, and more perish unseen out at sea. The growth of the Bering Sea fishing industry and heedless fishing practices are partly to blame, but much of the debris comes from farther off in Asia. Even one of the plastic tub toys, a green frog, washed up there—the northernmost siting from that spill.

The Pribilofs' plague reflects their role as entrapment islands; like Sable Island off Newfoundland and Matagorda off Texas, they sit alone, encircled by swirling currents. Alas, these same currents make many entrapment islands and collector beaches favorite haulouts and breeding grounds for marine animals such as the fur seals. Matagorda is one of the few beaches where highly endangered Kemp's Ridley sea turtles lay their eggs—clawing through the trash to find a sandy spot.

One collector beach, I'd long heard, was the most heavily junked of all: Junk Beach itself, as local beachcombers often call it, also known as Laeokamilo, "Point of Twisting Water." It lies on the west side of Kamilo Point, near the southern tip of Hawaii's Big Island—the southernmost point in the United States and probably the Polynesians' first landfall when they

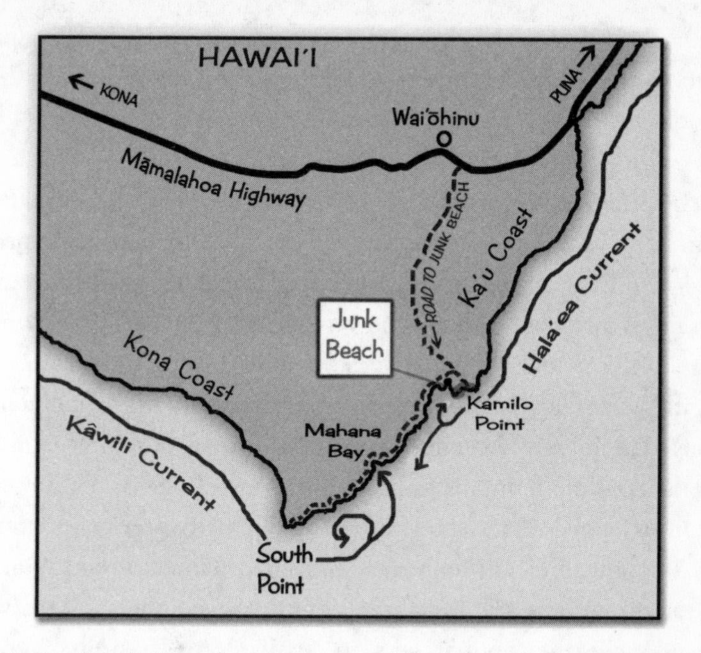

The road to Laeokamilo (a.k.a. Junk Beach), and the convergence of currents at Hawaii's South Point.

reached Hawaii. *Kamilo* means "swirling currents" in Hawaiian, and the name fits. It forms an elbow in the long, stark Ka'u Coast, which cuts west there and then turns south again—a sort of Malarrimo in reverse. The Hala'ea Current races down this coast from Puna, driven by the northeast trade winds. The rougher Kâwili (Hit and Twist) Current surges down the island's west side from Kona. They meet and merge at Kalae (South Point), five miles below Kamilo Point, creating a mini-gyre that sweeps flotsam into the natural catchment formed by the point—ergo, Junk Beach.

"Like the air flowing over the top of a pickup truck which collects lighter objects behind the cab" is how Captain Charles Moore, the commander of the marine-plastics research vessel *Alguita*, had described this shore to me. And, Moore added, though the seas off Kamilo Point are particularly rich fishing grounds, fishermen avoid them for a hundred miles out. They fear getting tangled in all the ghost nets and other trash that have accumulated there. That's because another powerful, if not exactly natural, phenomenon also converges here. This is the southwest corner of the North Pacific's Great Eastern Garbage Patch, the sprawl of sea into which any flotsam in the Turtle Gyre that doesn't wash up or escape on the conveyor eventually spins, and where it slowly gets ground to soggy dust.

For a certain adventurous sort of beachcomber, however, Junk Beach is a goldmine. And the queen of junk beachcombers is an artist named Noni Sanford who lives in Volcano Village, forty-two miles away as the seagull flies. Noni is a stocky woman with a ready smile, a long gray ponytail, a closet full of Hawaiian shirts, and a story to tell. Her father, Mark Lebuse, was a gifted artist, a swaggering bully of legendary strength, and a bit actor who played tough guys in gangster pictures, Asian action films, and *Hawaii Five-O*. Widowed but untamed, he caroused with Bogie and Tracy, learned to dive from Jacques Cousteau, brawled as much offscreen as on, and stayed a short step ahead of bill collectors and jilted flames. When Noni was a child, sometimes she and her three siblings tagged along on his wanderings. Sometimes they got parked with relatives or in an orphanage. Often they fended for themselves.

Such a seat-of-the-pants upbringing didn't allow much chance for formal education, and Noni attended twenty-eight schools, after a fashion, as she was growing up. But it left her avidly curious about the wide world, with a deep sympathy for ravaged people and places. When a stroke rendered her roguish father helpless, she took better care of him than he had of her. Today she serves as dispatcher at the volunteer fire department guarding the sprawling desolation of south Hawai'i Island, where the earth's inner fires ooze to the surface. And she makes art from the plastic scrap that Junk Beach so lavishly provides—not the usual cute driftwood lawn sculptures but stranger constructions, such as miniature replicas of her beloved yellow fire engine. Her art, she explains, "focuses on making something funny, useful, sad, or beautiful out of something that otherwise is just trash. Toothbrushes attracted me because of the ultimate ickiness that they represent."

On a Friday afternoon in January 2007, Noni drove around the island to the Kona-Kailua airport to pick up me and Dave Ingraham, Jim's son and my indefatigable photographer, videographer, and companion on many expeditions. We soon realized we'd made a big mistake flying into Kona rather than Hilo on the island's opposite side. A volunteer crew from Hilo was going to conduct a semiannual cleanup on Sunday, and I was scheduled to give them a pep talk and slide show that very evening.

Noni raced to get us across the Big Island, which meant crossing climate zones, from blazing sun over arid Kona to a downpour in lush Hilo. We arrived just in time for the talk. Afterward, the rain was still pouring when we reached her home in Volcano, close by steaming Kilauea Crater. We stayed up past midnight packing so we could make an early start, then rose at five to feast on Noni's crepes and set out for Junk Beach—hurrying to see it in its full ghastly glory before the Sunday cleanup.

The unmarked road to the beach cut off from the Māmalahoa Highway at the sleepy village of Wai'ōhinu. It was a road only in the most generous sense of the word: an unmaintained path that hurtled over powdery gravel for short stretches, then climbed over boulders and plunged into pits that would snare all but the fittest, shortest four-wheel drives. Your standard suburban land ship would be useless here; its long chassis would soon catch on the boulders.

Noni steered artfully in her red Jeep Wrangler. Dave rode behind with Noni's husband, Ron, in their orange Unimog, a looming, 22-speed, punishingly uncomfortable off-road juggernaut made by Mercedes Benz; in its lowest gear, it can inch forward too slowly for the human eye to detect. The Jeep made noises I'd never heard from a vehicle before, and I gave thanks that the Mog covered our tail.

The eight-mile road was one capillary in a matrix of trails that ranchers and fishermen had cut over the centuries through the blind thickets. It took an hour to negotiate it; even after many trips, Noni took wrong turns. The thickets finally opened up onto a Mars-like lava plain. The boulders grew bigger and the crevices deeper. The Jeep slowed from a crawl to a creep.

Around us, lava-rock walls crisscrossed the plain like dry-stone walls in New England. Noni explained that ancient Hawaiians had laid them to demarcate the territories of various royals, back when this desolate landscape was the most densely settled part of the island. It was hard to imagine bare-footed, bare-handed laborers slinging the jagged, siliceous volcanic rocks into place. I hoped the Jeep's tires would prove as resilient, and wondered why anyone would work so hard to secure such a barren, blasted landscape. Agriculture was one reason, Noni explained: Once fresh water lay close to the surface, and taro and other crops thrived in the mineral-rich volcanic soil. Then cattlemen came and planted Mediterranean algarroba trees, whose sweet carob pods provided rich fodder. The algarroba sucked up the water, driving off the people who had prospered here for centuries.

I guessed another reason this windswept territory would have been so valuable: the flotsam that accumulated on Junk Beach. For the ancient Hawaiians, with their endless internecine and inter-island wars, seafaring dugout canoes—catamarans and twin-hulled *peleleu*—spelled power and conquest. The larger they were, the more warriors they carried, a decisive advantage in close combat; full-size canoes were a hundred feet long, more or less matching Columbus's ships. In 1795, Kamehameha, the first king to rule all the islands, assembled an armada of five hundred canoes to attack Maui and Oahu to the north.

For all their lush forests, the islands were, like Iceland, sadly deficient in the tall, straight-trunked trees needed to make such watercraft. But the gods offered a gift: enormous cedar, spruce, fir, and redwood logs from mainland forests two to three thousand miles away. These would periodically wash down the rivers there, out to sea and, eventually, onto collection points like Kamilo. Modern Ka'u residents made surf boards from them. Washed-up "fir timber . . . is by no means uncommon," observed Lieutenant George Vancouver, who sailed with Cook in 1793, "especially at Kauai, where there then was a double canoe, of middling size, made from two small pine-trees, that were driven on the shore nearly at the same spot."

The value of these precious logs comes clear in a tale recorded by the nineteenth-century missionary Titus Coan. One of his colleagues was trying to translate Paul's admonition "Add to your faith knowledge, and to your knowledge temperance, and to your temperance virtue" into Hawaiian. "He appealed to his native assistant for the Hawaiian word for virtue, which he described as the most desirable of all possessions. The native was puzzled; neither the conception of virtue, as we understand it, nor any corresponding word, existed in Hawaiian; but at last he said: 'I understand you now,' and gave the missionary a word which made the passage read: 'Add to your faith knowledge, and to your knowledge temperance, and to temperance a stick of Oregon pine.' "

For years, I'd corresponded with the distinguished Hawaii historian Ruth Levin. On this trip we finally met in person, at the Hawaii Volcanoes National Park Museum. She told me that when tsunamis and big floods hit, logs would wash far upland onto the lava plains. And floods on the mainland would send bumper crops of logs Hawaii's way. In the winter of 1861–62, a succession of snowy Arctic blasts and wet tropical storms inflicted a thousand-year flood on California, inundating the Sacramento River Valley to a depth of ten to twenty feet. Inland waves smashed farmhouses to pieces and bore vast rafts of toppled trees out to sea. About a year later, Hawaiians reported a horizontal forest floating past. Perhaps they called it Tree Beach rather than Junk Beach.

William H. Brewer, president of the National Academy of Sciences, later visited Crescent City, farther north on the California coast, where the Davidson Current had transported some of the flood debris. There he saw logs "in quantities that stagger belief . . . great piles, often half a mile long, and the size of some of these logs is tremendous." Brewer measured one that was 210 feet long and three and a half feet thick "at the little end, without the bark."

Today the evening news reports excitedly on all the houses, cars, and other flotsam washed away in floods. Rarely, however, do we learn what happens afterward to this diluvial debris. Some of the trees washed away in the great 1861–62 flood stranded on nearby shores. Coastal eddies, observable from earth-orbiting satellites, spun others a hundred miles offshore, where the California Current swept them on westward to the Hawaiian Islands. In September 1862, Charles Wolcott Brooks, secretary of the California Academy of Sciences, reported "an enormous Oregon tree about 150 feet in length and fully six feet in diameter above the butt" drifting past Maui. "The roots, which rose ten feet out of water, would span about 25 feet. Two branches rose perpendicularly 20 to 25 feet. Several tons of clayish earth were embedded among the roots"—carrying who knows what biological invaders to vulnerable island habitats.

Any logs that got past Hawaii without being snatched or washed up would, over the next five to ten years, complete a full orbit around the Turtle and/or Aleut gyres. As we jostled down the ruts and rocks to Junk Beach, I imagined how such trees, the keys to wealth and war, might have transformed societies all around the North Pacific.

Like the Norse on Iceland, the Hawaiians settled where the driftwood pickings were good. Their kings placed sentries along the windswept southern coast to watch for big trunks floating offshore, and when one appeared, canoes would scramble to drag it in. Noni and I saw the cliff-top ruins of old sentry quarters, including one near Junk Beach. I realized that the lava walls we'd noticed coming down testified to the power of driftwood, which had drawn these people here. Even their name for this wind-blasted shoreline showed how rich and generous they thought it to be: Ka'u means "the breast."

Ruth Levin shared other lore that suggested just how much the Ka'u dwellers esteemed the powerful offshore currents. Once they used these currents to end the rule of a greedy chief named Hala'ea, the namesake for the current that sweeps down from Puna. Each evening Chief Hala'ea paddled out to meet the returning fishermen and exercise his royal rights to the best of the catch. Often he took it all, wantonly wasting food while the commoners went hungry. At last, the fishermen collaborated to pay him back. When Hala'ea paddled out and called, "Give me the fish," they loaded so many onto his canoe that it sank and the current swept him away.

When Ka'u Coast dwellers lost loved ones at sea, they would seek their bodies on two separate stretches of beach bracketing Kamilo Point: Ka-Milo-Pae-Ali'i, "the twisting water washes ashore royalty," and Ka-Milo-Pae-Kanaka, "the twisting water washes commoners ashore." The fat-rich bodies of kings and chiefs and their lean subjects washed up at different sites; even in death, the beaches were divided by class. I could not help thinking of the medieval trials by water.

The currents off Kamilo were also said to serve a happier use, as a sort of waterborne postal service. According to the local historian Mary Kawena Puku'i, travelers to the adjacent Puna district would cast leis tied with loincloths and pandanus leaves into the sea when they arrived. When these floated back to Kamilo, the senders' families knew they had arrived safely.

In 2003 and 2005, I tried to replicate this early use of drifters with biodegradable drift cards—thin plywood painted bright orange—that I'd devised years earlier to study the fate of sewage in Puget Sound. The cards bear a phone number and an inscription asking anyone finding them to report where they ended up and when. Ruth Levin threw thirty-five cards off the cliffs (and into the wind) at Puna; only one showed up at Kamilo, seven days later. Two out of twenty-five that she arranged for a fisherman to cast further offshore washed up at Kamilo, thirteen days later. She suspects the cards were thrown too close to shore and got trapped by the surf; she wants to row farther out and cast another batch safely past the surf. For now, the Puna-to-Ka'u current seems to have been a rather unreliable postal system.

But Kamilo remains a dreadfully efficient flotsam collector. Most of what washes up now is very different from the logs, seashells, and other natural materials of the old days. Our little party reached Junk Beach in early afternoon. It was a classic Hawaiian beach, a mile of bright, coarse sand facing a gentle shelf of sand and lava rock ringed by offshore breakers. But its entire length was strewn with rubble, piled in drifts two feet thick: ghost nets and ropes, driftwood and scraps of paneling, and, especially, plastic—every size, shape, and color, from whole barrels to tiny specks.

Dave, Noni, and I beachcombed for much of the day. The sun blazed, piercing our number 30 sunblock and giving Dave a nasty burn. The trade winds blew as they usually do here, twenty to thirty miles an hour. A sudden shower drenched us, and I felt briefly as though I were back in Seattle. Then the sun returned and dried us.

That sun was the ultimate cause of all that we saw. It heated the air that rose, fell, and circled about, creating the garbage patch. It also, less directly, created the plastic that now filled the patch and covered Junk Beach. Mil-

Curt (left) and Noni Sanford gaze appalled at transoceanic debris on Hawaii's Junk Beach.

lions of years ago it nourished the plants and animals that decomposed and formed the petroleum from which we make plastic.

Four years earlier, Junk Beach's junk rose much higher—seven, eight, even ten feet, according to local residents. Then, in 2003, the State of Hawaii funded a cleanup. Volunteers removed fifty tons of nets and other dangerous marine debris. In September 2004, the U.S. Geological Survey sponsored another cleanup. The high-seas plastic patroller Charlie Moore, who joined in the cleaning, conducted a revealing inventory. In a square-foot patch of shore midway along the high-tide strand line, he collected twenty-five hundred plastic fragments a millimeter or more wide. Five hundred of these were BB-sized pellets, the form in which raw plastic is shipped around the world and spilled into the sea when containers topple. California lifeguards, watching these pellets wash up on their beaches, gave them a name that's stuck: "nurdles." The Kamilo cleaners did not touch these or millions of other small plastic chips; they took only debris that was fist-sized or larger.

In the winter of 2005–6, the Hawai'i Wildlife Fund and the National Oceanic and Atmospheric Administration sponsored another, more extensive cleanup. Several large crews of volunteers cleaned the nine miles of beach between South Point and Wai'ōhinu. They removed thirty-six tons of fishing line and nets—lethal ghosts that, if washed back out by storms, could strangle turtles, fish, birds, and the endangered monk seals that haul out on these shores—and shipped them to Honolulu for burning in a trash-to-energy generating plant. And they hauled another six tons of plastic, glass bottles, and other trash to local landfills.

More debris washed up even as the crews finished cleaning. A year later, when Dave and I arrived, an estimated fifteen to twenty tons had accumulated. Scattered amid the litter were balled-up nets and industrial bales of Velcro one to two feet thick. A boulder-like block of rubber latex, more than two feet square and dark and pocked like the upland lava rocks, lay among the litter. A cloud of honeybees lapped at its gummy surface.

At the shore's edge, the water was a stewpot of tiny plastic chips, rippling with the waves and wash-back. I could not help thinking of a bizarre party, or parade; the sea was showering confetti on us, saluting the world we'd made by throwing bits of it back at us. White and baby blue were the litter's predominant colors; other hues fade faster in the tropic sun. When the water became calm, the confetti would rise to the surface, coating it for five to ten yards out, so solid to the eye that you could imagine walking on it. Then a splash would hurl it up on the shore, mixing it with the sand and wood chips. It formed parallel tide lines and filigree outlines—here a map of Eurasia, there a chubby Lascaux-style horse. I counted seven tide lines of confetti, highlighted by the lowering afternoon sun.

"The tides of plastic," Dave groaned. "It makes me ill."

"It looks like vomit," I replied, and realized that it was just that: the disgorgement of an ocean overstuffed with human detritus.

Much more of the plastic in the garbage patch never leaks onto Junk Beach, but also never sinks or disintegrates entirely. Instead it floats until it crumbles or gets swallowed; plastic molecules are remarkably persistent.

You might think that this plastic invasion has been raging over Junk Beach and Hawaii's other wash-up beaches ever since the plastic age got into full swing in the mid-twentieth century. But according to beachcomber James Marcus of Waimanalo, Oahu, large volumes of pulverized plastic only began beaching on Oahu's windward side in February 1998. He sent me samples of this plastic sand, and though it appeared to be quite fine, it was actually much coarser—about seventy-five hundred bits per pound—than the hundred thousand pieces per pound that oceanographers have obtained trolling the North Pacific. Since then, plastic sand has swept onto all the islands' windward shores, along with shoes from the 1990 Nike spill, tub toys from the 1992 spill, and hockey gear from the 1994 spill.

It appears that the Eastern Garbage Patch released a huge batch of flotsam all at once, perhaps because the high-pressure cell above it temporarily abated and the winds rose. Might this be a recurrent, even cyclical phenomenon? Might such releases bear any correlation to the lunar node and Pacific Decadal Oscillation? OSCURS simulations have shown that flotsam can swirl around a gyre as long as sixty years, breaking into smaller and smaller pieces under the beating of waves, surf, and sun. But Jim and I have not had the opportunity to examine this possible linkage. We have much yet to learn about the gyres and their innards, the garbage patches.

You load sixteen tons, whaddaya get? I muttered, staring at Junk Beach and trying to suppress my disgust and amazement. I recalled the rule of thumb I'd derived from years of combing other beaches: every three tons of trash contains a nugget of scientific gold—a survey stake or another artifact traceable to its place and time of origin. I examined the junk strewn along Junk Beach. Jim Ingraham and I had long been projecting the movements of flotsam around the North Pacific, extrapolating out as far as sixty years. These projections suggested that a large, though still undetermined, share of the garbage in the patch originated in Japan.

Studies done in the seas off Japan, which feed the Turtle Gyre and the Great Pacific Garbage Patch, comport with that finding; they show plastic particles increasing tenfold every ten years in the 1970s and 80s—and tenfold in just three years in the 90s.

As we have seen, ocean memory has a half-life; a gyre sheds about half its debris with each orbit, six years in the case of the Turtle Gyre. But the particles are constantly breaking down, and thus becoming more numerous, as they orbit—an unending cascade to a virtually infinite number of microparticles and molecules.

A constant stream of new debris is always washing in from rivers, ships, sewers, floods, landfills, and urban runoff. The paper in it will dissolve rapidly, the wood, iron, and fabric a little more slowly. No one knows how long petroleum-based plastics will hold up; estimates range from five hundred to a thousand years. On this score, Charlie Moore and I see things somewhat differently. In 2003, he cautioned that it was "a bit of fraud" to suggest that beachcombers might still find whole tub toys rounding the gyres or the great conveyor after eleven years; more likely, they'd have crumbled into pieces. But that same year, at the Sitka fair, one beachcomber showed me a red beaver, faded nearly white but still intact, which he said had just washed in.

Since I assumed custodianship of the drift-card archive that NOAA abandoned, I've been receiving older and older cards that have washed out of the gyres. In May 2008 I received one that was released off Nantucket in 1976 and had just washed up in Spain; after thirty-two years at sea, the printing on it was still legible. At this rate, a gyre cannot purge itself; it will carry plastic forever.

Research elsewhere shows an even more dramatic increase in marine trashing. In the Southern Ocean, which embraces the great Penguin Gyre, the amount of debris ingested by Patagonian petrels increased a hundredfold in the same decade. Charlie Moore and his colleagues in the Algalita Marine Research Foundation surveyed a globe-spanning range of flotsam inventories and found that plastic typically made up 60 to 80 percent, and sometimes more than 90 percent, of total debris. Towing nets from the *Alguita* in the central Pacific, they found six times more plastic than plankton by weight.

The various plastics have particular densities. High- and low-density polyethylene, polypropylene, and polystyrene foam (recycling codes 2, 4, and 5 respectively) float. Polyethylene terephthalate, polyvinyl chloride, and solid polystyrene (codes 1, 3, and 6) sink. The former represent about 46 percent of plastic sales, so about the half of plastic goods, like half of human bodies, are floaters. This also means that plastic debris is found at all depths: floaters on the ocean surface, neutrally buoyant objects throughout the water column, and sinkers in sediments both deep and shallow.

The past few years have brought a flurry of attention to the plague of plastic flotsam, thanks in good part to Charlie Moore's efforts to publicize it; nothing seizes the media's and the public's attention like a maverick, crusading sailor-turned-scientist, except perhaps a bathtub duck lost at sea. But despite the endlessly rekindled novelty of the subject, there's really nothing new about it. Scientists have investigated plastic dumping and its

Captain Charlie Moore, tireless crusader against ocean dumping and plastic pollution, with prizes hauled from the Eastern North Pacific Garbage Patch.

marine impacts since the 1970s. In the 1980s, scares over floating trash, in particular syringes and other medical wastes, closed a hundred New York–area beaches, costing beach-town economies billions of dollars. In 1988, the nations of the world came together to try to correct it with the International Convention for the Prevention of Pollution from Ships, commonly known as MARPOL (for "marine pollution"), a treaty banning the dumping of all plastic and most other garbage at sea. By 2005, 122 nations had ratified MARPOL. Some studies suggest that MARPOL has reduced both marine debris and entanglements by derelict fishing nets in some places, notably the Alaskan and Californian coasts. Other studies show no improvement in other areas, including the Southern Ocean, South Atlantic, and Hawaiian Islands. As with so many environmental protections and international treaties, enforcement lags far behind stated intentions. And though an annex to MARPOL requires that signatories provide shoreside facilities where ships can dispose of their garbage, many developing countries have not been able to do so. Even captains and shippers who want to comply may not be able to.

And even if, with unerring enforcement and perfect compliance, MARPOL miraculously eliminated ocean dumping, it would only slightly lighten the sea's burden of trash. The *Alguita*'s net trawls suggest that fully 80 percent of marine debris comes from land. Greenpeace estimates that 10 percent of the 100 million tons of plastic produced each year worldwide ends up in the sea. That global production includes, by various estimates, 500 billion to 1 trillion plastic bags. It takes just one bag to choke a hungry sea turtle. If that 10 percent estimate holds for bags, then enough drift into the sea each year to kill all the sea turtles in the world thousands of times over. One shipping container holds about 5 million plastic bags, and I know of at least two such containers lost in Turtle Gyre. No one knows what happened to their 10 million bags. The shipping industry is proud that it's reduced its annual loss rate from about ten thousand to two thousand containers out of roughly 100 million shipped each year. I tell them it only takes one to cause a catastrophe.

Extrapolating from a sample area, I estimated that twenty thousand of the plastic cones used to catch hagfish—primitive, eel-like deepwater fish that are delicacies in Japan and Korea—lined Junk Beach's one-mile length, along with twenty thousand oyster-spacer pipes, pencil-length plastic tubes used to separate the scallop shells on which aquacultural oysters are grown. With them we found half a dozen Japanese survey stakes—heavy, foot-long plastic shafts that occasionally bear imprints, which, translated and traced, provide valuable information as to how long they've traveled and what routes they've taken. I found four "Creap caps"—lids from Japan's answer to Coffee Mate, its name a conflation of the English words "creamy powder"—uncountable plastic fishing floats and octopus traps, and thirty toy wheels, which, I must confess, I enjoy collecting. They remind me of the oceanic gyres.

Even as it confirmed our predictions, however, Junk Beach overwhelmed me. Despite all our prior computer modeling, I gasped at the sheer concentration of plastic detritus; for a flotsamologist, it was like trying to drink out of a fire hose. Dazed and appalled, I forgot to gather samples of the plastic confetti as I'd planned. I wanted to determine how the confetti affects the reflective qualities of the water in which it floats, a question with serious implications in a warming world.

That night Dave grilled tuna steaks over briquettes ignited with a blow torch. We washed our tuna down with beer and, wrapped in the tropic night, forgot the now-invisible devastation around us. The wind came up, and heavy rain followed. Ron and Noni stretched a tarpaulin fence to fend off the wind, and more tarps over the bed of the Unimog to make dry sleeping quarters.

Sunday morning still blew squally. Noni and I beachcombed while Dave and Ron, using the Mog as a tractor, dragged ten ghost nets up past the reach of the tides, where they could no longer kill. The squalls now blew through every hour or so.

By noon it was clear that the planned cleanup—the reason we'd rushed to Junk Beach the day before—would not happen today. So we took a side trip along the rugged Ka'u Coast, the longest stretch of undeveloped shoreline left on the island. Mainlanders had tried and failed to exploit this shoreline as successfully as the Hawaiians had. From the mid-nineteenth century through the twentieth, cattle and sugarcane from upslope operations were loaded onto waiting ships from rough landings built of lava rock, but the last sugar plantation closed in 1996 and the cattle are now shipped off by truck. Proposals for other developments—resorts, prisons, even spaceports—have arisen and sunk.

The Kamilo area is now protected; the state has dedicated it to wildlife conservation and archaeological preservation rather than cattle grazing. Local residents bounce down over the lava to fish, camp, and enjoy the coast's roughhewn beauty. Endangered monk seals raise their pups along the shoreline among rare, endemic coastal plants. Turtles and seabirds flock here.

On the way to South Point, Ron and Noni pulled off the road to show us a small inlet called Awawaloa (Too Salty), perhaps fifty yards wide and a few hundred yards long, flanked by sheer lava cliffs. Mighty waves crashed furiously here, stirring a maelstrom that seemed fit to smash any object or creature that strayed into it. Two enormous turtles bobbed amid the ten-foot surf, feasting on the algae that grows profusely in such hyperoxygenated conditions. But their situation was precarious. Beside them, several plastic ghost nets also bobbed—a strangling death even for mighty turtles unlucky enough to get entangled.

The cleanup crews arrived at Kamilo Point a few weeks later, after the squalls blew over. But they did not remove the confetti and other small plastic remnants; that would remain for the tides to reclaim and the gyre to grind down further. The sea will remember this for the next century, I mused; long after the world runs out of oil, some of this scrap will still be circling the ocean. My skin crawled. Junk Beach was a haunted place—haunted by the ghosts of plastic past, and premonitions of the future.

II. The Synthetic Sea

It ate the food it ne'er had eat,
And round and round it flew.

—Samuel Taylor Coleridge,
"The Rime of the Ancient Mariner"

Plastic is the great pretender, conceived in mimicry and dedicated to the simulation of everything else in art and nature. Simulation drove even the creation of the first industrial plastic. In the mid-nineteenth century, two fixtures became essential accessories of the good life as America and Europe grew wealthier and more status conscious. For men, there was the pool or billiard table, the centerpiece wherever gentle and not-so-gentle men gathered, from corner saloons and barbershops to millionaires' mansions; mules even bore the heavy tables up to the Yukon and Klondike goldfields. Women, children, and families meanwhile gathered around the parlor piano, tickling one set of ivories while men trained their cues on the other.

Together these formed the Victorian home entertainment center. And they shared one other feature. They depended on the same exotic, pricey, and fast-shrinking resource. Ivory provided just the right grip and moderate hardness for pianists' fingers; it even absorbed the sweat off them. And only ivory afforded the liveliness, elasticity, and "click" pool players craved. The rush to supply them—billiards in particular, by far the greedier consumer—

ravaged Africa's elephant herds and human societies. Hunters and traders scoured ever-wider swathes for the "elephant teeth" that had once been so plentiful they were left to rot. The annual toll topped one hundred thousand; in 1867, the explorer David Livingstone estimated that "44,000 elephants, large and small, must be killed to supply the ivory which comes to England alone."

Prices soared; Brunswick charged $7.50—a skilled worker's weekly wage—for a clear, select carom ball. But the ball makers worried about shrinking supplies. In the early 1860s, as the thinly documented story has come down, another leading manufacturer, Phelan & Collender, offered $10,000 to anyone who could devise a suitable ivory substitute, sending a generation of tinkerers racing to their workshops. Among them was an ex-seminarian, self-taught jack-of-all-trades, and printer-turned-inventor named John Wesley Hyatt.

Hyatt turned to a new compound called nitrocellulose, created in 1846 by treating plant cellulose with nitric and sulfuric acids, as his base. England's Alexander Parkes had been trying for years to concoct a moldable plastic by mixing nitrocellulose with various oils, but the results bent, warped, and, given half a chance, burst into flames. Hyatt formed billiard balls by coating spheres of pressed wood and shellac with dissolved nitrocellulose, but the coating shrank and bubbled as it dried. Then he hit upon a solution. Eschewing oils and solvents, he ground up nitrocellulose with the aromatic crystal camphor—extracted from Asian trees that are often found floating in the Pacific—and subjected the mix to high heat and pressure. The result was a hard, transparent block, or ball, or whatever shape he chose to mold. In camphor, Hyatt had discovered the first industrial plasticizer—a material that makes another material soft and pliable. And he had created the first successful thermoplastic.

Hyatt dubbed his new substance "celluloid"—cellulose-like. The mimicry had begun. Celluloid could be shaped and tinted to imitate not just ivory but shell, bone, glass, wood, gems, marble, pearl, and cloth. It was molded into everything from hairbrushes to sculpted flowers. But celluloid pool balls never caught on in a big way, though they were sold until the 1960s. Hyatt instead scored his first big celluloid success with denture plates, followed by collars and cuffs. Then, in the 1880s, a manufacturer of glass photographic plates named George Eastman got the idea of making flexible, multiple-exposure celluloid film. This transformed photography from an elite, awkward art into an inexpensive popular medium—and put the magic of reproduction in the hands of the masses. Celluloid film begat motion pictures—simulation in four dimensions—and led, indirectly, to television and all the digital media that have followed. Everyone who sees a movie or shoots a snapshot has John Wesley Hyatt to thank.

But already plastic was showing its perilous side. Nitrocellulose and its derivatives had a nasty tendency to burn or blow up. The nitrocellulose-coated pool balls Hyatt initially sold would sometimes burst into flames when a cigar passed too close. Or they would, as he put it, "produce a mild explosion like a percussion guncap" when they struck each other. One Colorado saloonkeeper wrote Hyatt that he wouldn't mind these little blasts, "but that instantly every man in the room pulled a gun."

Hyatt's celluloid was nowhere near as incendiary as the explosive form of nitrocellulose known as guncotton, but newspaper wags never let such fine distinctions get in the way of a good column. In 1875, the *New York Times* spun a tale of a Vermont swain who gave his sweetheart stainproof, wrinkle-free celluloid cuffs and collar; when he lit a cigar, she vanished in a terrific blast. A less-fanciful editorial reassured consumers that their celluloid items would *not* explode; they'd just burn like torches. Likewise, the so-called nitrate film that Eastman developed from Hyatt's celluloid; projection rooms were lined with asbestos, and showing and storing movies was notoriously hazardous until more stable acetate film replaced celluloid in the 1950s.

Celluloid pool balls were superseded much sooner. In 1909, the Belgian-American chemist Leo Baekeland unveiled Bakelite, the first cast phenolic resin and first truly synthetic plastic—harder, tougher, and much less flammable than celluloid. Since then the pace of innovation and the volume of manufacturing have multiplied, as plastics substitute for more and more materials in more and more uses.

But plastic's gifts of impersonation go much deeper than that. At each stage of its breakdown, plastic in the ocean uncannily mimics the organisms—anchovies, copepods, and finally phytoplankton—that other organisms feed on. Ingested, it travels up the food chain, concentrating in larger, longer-lived animals. Imagine salting your food with plastic ground to dust. That is what's happening to the oceanic food chain, all the way up to us.

You can hardly blame seabirds and turtles for mistaking plastic fragments and pellets for food; researchers in oceanographic laboratories have misidentified nurdles as fish eggs. Albatrosses, scooping up whatever glimmers on their long glides, are especially vulnerable. A 1969 examination of one hundred Laysan albatross carcasses beached in the Hawaiian chain found an average of eight indigestible items in each stomach—about 70 percent pumice and 30 percent plastic fragments. Three decades later plastic predominated. In 2005, the nature photographer David Liittschwager asked me to examine a shot his colleague Susan Mittleton had taken of the

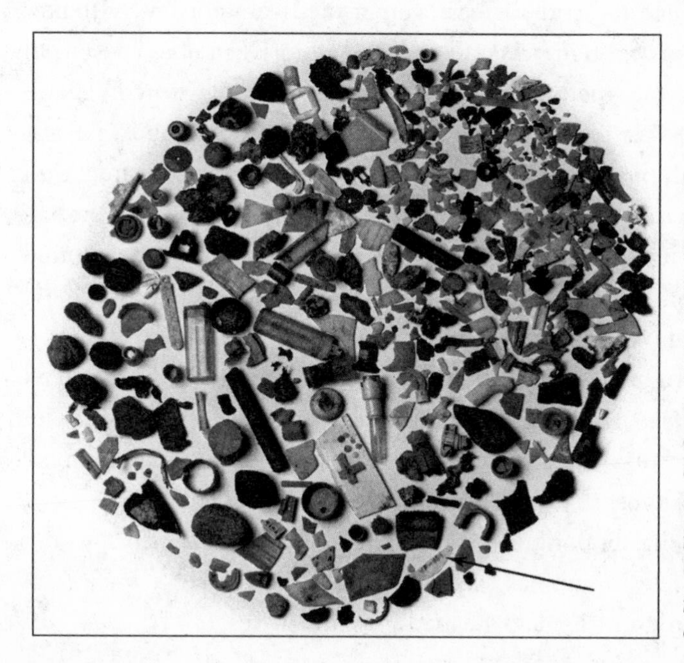

More than five hundred pieces of flotsam were found in the digestive tract of a dead albatross chick on Midway Island at the northwest end of the Hawaiian chain. At lower right, the VP-101 plastic tag was traced to a naval patrol bomber downed in 1944.

stomach contents of an albatross chick found dead on Midway at the chain's north end. It had swallowed more than five hundred pieces of debris, including sea beans, Bic lighters, shotgun-shell cups, toy wheels, and the plastic tubes used as spacers in Japanese oyster farming. (Short and long tubes both wash up on Japan, but it is mostly the long ones that make it across the Pacific. Perhaps albatrosses swallow the short ones in transit.)

Most of the items were so badly degraded I could not identify them. But one told a remarkable story—and set a new survival record for intact plastic objects at sea. It was a half-inch tag stamped "VP-101." After *National Geographic* published the photo in its October 2005 issue, a reader wrote to say that a U.S. Navy aerial patrol squadron that served over the Pacific in World War II was designated VP-101. "Is it likely that this item has been floating in the Pacific for over sixty years?" a researcher at the magazine asked me.

Until then, the longest-floating objects I'd been able to date were a 1955 glass ball and a rubber ball decorated with early 1950s Warner Bros. cartoon characters, both of which washed up in 2003. I contacted various veterans from that era, including one retired commander who was writing a history of the VP-101 squadron, and pieced together the likely story. VP-101 flew amphibious PBY patrol bombers and operated from December 1940 until 1943 or '44. The tag—perhaps made of durable Bakelite—wouldn't have adorned an actual plane, but it might well have labeled a toolbox, navigation sight, or other piece of gear. VP-101 lost planes at seven known sites off the Philippines, Indonesia, and Australia; from the Philippines, the tag could have been drawn into the Kuroshio current, carried nine times around the Turtle Gyre, and eventually flashed before a cruising albatross, whose sharp eyes spotted what a human beachcomber would likely overlook.

And here was a final grace note. As a child in California I would marvel at PBY "flying boats" gliding as smoothly and, it seemed, effortlessly as an albatross over the surf. Bird and plane, both amphibious, hunted their prey in the same way, across the same five-thousand-mile range.

Even in the ghastly poisoning of a magnificent bird, the sea spits out elegant symmetries. Sometimes I feel like an albatross myself, choking on so much grim but exquisite data gleaned from the waves.

Plastics continue their deadly mimicry right down to the molecular level. Many of the chemicals used in modern plastics mimic hormones, in particular the female hormone estrogen, disrupting fundamental reproductive and physiological processes in humans and other organisms. Researchers and representatives of the plastics and chemical industries still dispute how much these "endocrine disruptors" actually affect health and how much is assuredly known on that score. But more evidence accumulates every year about the seriousness of these effects, about new chemicals that function as endocrine disruptors, and about their pervasiveness in the seas of the world.

Many of these disruptors are xenoestrogens, external substances that bond to the body's receptors for estradiol, the natural estrogen that triggers estrus, lactation, and the development of genitalia and other female sexual characteristics. In females these extra estrogenic bursts, without the metabolic outlet of reproduction, can encourage various cancers. The effects may not be confined to—and may not even strike—the generation that actually takes in the chemicals. The first synthetic estrogen, DES (diethylstilbestrol), was widely used to prevent miscarriages from the 1950s to the 1970s. (It has also been used to treat breast cancer and fatten livestock, and for hormone-replacement therapy.) Many of the daughters of mothers who took DES were found to have deformed or underdeveloped reproductive organs and various cancers, including a rare vaginal cancer; this was the first known instance of a cancer-causing substance crossing the placenta. Though attention naturally focused on these horrors, male children of DES-taking mothers weren't exempt. Many of them suffered corollary ills and deformities: testicular cancer, undescended testes, extremely small penises, exposed urethras.

Around the same time, two other widely used chemicals became notorious for their estrogen-like qualities: DDT (dichloro-diphenyl-trichloroethane), the miracle insecticide of the 1940s and 50s, and superstable PCBs (polychlorinated biphenyls), which were widely used as electrical insulators and in wiring, paints, caulking, hydraulic oils, carbonless copy paper, and a host of other products.

Alarms rose over this terrible trio in the 1960s and 70s. The industrial nations banned PCBs, which also cause a wide range of other toxic effects, including anemia, liver cancer, and neurological damage. But because they are so stable, they continue to leach out of landfills, industrial sites, and harbor sediments, traveling up the food chain and poisoning generation after generation. The PCB-laden killer whales that are Puget Sound's peak predators carry the heaviest toxic loads of any vertebrates ever measured.

DDT devastated songbird and raptor populations by causing them to lay thin-shelled eggs doomed never to hatch. This avian holocaust inspired Rachel Carson's best-selling exposé *Silent Spring,* which in turn inspired both the modern environmental movement and a U.S. ban on DDT (though it's still used in many malaria-afflicted nations). Since then, other pesticides and industrial chemicals—endosulfan, methoxychlor, heptachlor, toxaphene, dieldrin, lindane, atrazine—and even the toxic natural metal cadmium have been shown to bind to estrogen receptors. Others, including dioxin, lead, and, once again, PCBs, disrupt another part of the endocrine system, blocking the thyroid gland's production of hormones that are essential to growth, metabolism, and reproduction.

In the 1990s, researchers in America and Europe observed a *Rocky Horror Picture Show*'s worth of gender-bending, procreation-sabotaging abnormalities. One of the highly polluted St. Lawrence River's fast-dying belugas had nearly complete sets of male and female reproductive apparatus—the first hermaphroditic whale ever reported. Like the belugas, seals in the Baltic Sea and along the Dutch coast showed high rates of reproductive failure, in step with their PCB levels. Mink along the Great Lakes and otters in England accumulated the PCBs in their fishy diets—and died out.

In one signal case, highly endangered Florida panthers showed weakened immunities, thyroid malfunction, and high rates of sterility in both sexes, plus low sperm counts and undescended testicles in the males. For decades authorities blamed inbreeding and even imported Western cougars to enrich the gene pool. Then another likely cause emerged: the pesticide soup washing down from vegetable farms, fruit groves, and sugar plantations in the slow-moving river that is the Everglades.

Humans, not surprisingly, are not immune to such effects. Various researchers have reported falling fertility and rising numbers of genetic deformities in the general population, not just DES babies. One much-noted but controversial 1992 Danish study found that average sperm counts had fallen 50 percent over the last half century; other studies have reported similar, smaller, or no declines. Where data is gathered seems to make a big difference; average sperm counts vary greatly from region to region, even within the United States. One reanalysis found that sperm counts had indeed declined significantly in Europe and North America, but not in other parts of the world. This might reflect more exposure to synthetic chemicals in the most industrialized nations.

Amid all these debates, a once-exotic term gained currency as a catch-all for these estrogen-induced alterations: "feminization." The whole world seemed to be undergoing a partial sex change—at the mini- as well as megafauna end of the spectrum. The tiny planktonic crustaceans called copedods—perhaps the largest biomass on earth—are an essential food for salmon and many other fish and whales. In 2008, researchers at Masaryk University in the Czech Republic reported that copedods are susceptible to some of the same synthetic endocrine disruptors as vertebrates; they likewise show altered sex ratios and high embryonic die-off.

For decades, plastics largely escaped scrutiny, even as they grew ever more ubiquitous. After all, they were solid, intact, wonderfully stable materials, not liquids that could leak from electrical transformers as PCBs do or, like DDT, get sprayed over entire landscapes. But warnings were sounded. In her 1997 book *Altering Eden: The Feminization of Nature*, the science writer Deborah Cadbury recounts how, clear back in 1936, Charles Dodds, the British chemist who created DES, reported that it was just one of a large class of chemicals called diphenyls or biphenyls that acted like estrogen. These include PCBs and their chemical cousins PBDEs (polybrominated diphenyl esters), which are widely used as flame retardants and in plastics, circuit boards, and synthetic fibers. PDBEs can compose up to 15 percent of a television casing and 27 percent of an upholstery fabric by weight. They persist like PCBs and enter the environment even more readily, via air and water. Since they started showing up in breast milk, several states and European countries have banned some PBDEs. But the industries using them insist they're necessary to keep their products safe.

Recently, two other common plastic additives have become chemical enemies number one and two in the public's mind. One, a diphenyl called bisphenol A, is widely used in epoxies, food-can linings, and to harden tough plastics such as PVC and polycarbonate, the stuff of Nalgene bottles, plastic baby bottles, and compact disks. The other, a group of chemicals called phthalates, serves the opposite purpose: they lend flexibility and squeezability to vinyl products such as baby toys, shower curtains, food containers, caulk, flooring, fishing lures, and "jelly rubber" sex toys. Some vendors advise using condoms with these toys to avoid toxic exposure.

Bisphenol A plastics were thought to be stable and benign until the early 1980s, when Stanford researchers made a bizarre accidental discovery. They'd attempted to wind the evolutionary clock back and determine whether even single-celled yeasts produced estrogen. Yes indeed, they found that the yeast hormone even seemed to be the same as human estradiol. But then they got an estrogen response with no yeast present. Something else was binding with the estrogen receptors in their cultures. Eventually they tracked down the culprit: bisphenol A leaking from the polycarbonate used to sterilize water for the cultures.

Human data is lacking for bisphenol A, but animal studies associate it with breast cancer, enlarged and perhaps more cancer-prone prostates, decreased testosterone, diminished maternal behavior, lower birth weights, and birth defects.

Though they're used to soften rather than harden plastics, phthalates seemed to behave like bisphenol A in one way. They're associated with undescended testes and male genital irregularities (along with liver damage and other problems). Phthalates were also widely viewed as xenoestrogens, until a closer look revealed that they're actually androgen antagonists. Instead of acting as female sex hormones, they suppress the production of male hormones. And the same slippery quality that makes them such

Plastic confetti, "plankton" in the making, blankets the water near Kamilo Point, Hawaii.

handy plasticizers—phthalate molecules slide loosely, like ball bearings, between plastic molecules—makes them leach out readily into the environment.

Phthalates and bisphenol A are ubiquitous in our consumer environment and our bodies. But how prevalent are the plastics that contain them in the garbage patches? There's painfully little data to go on. In 1999, on their first cruise in the Eastern North Pacific Garbage Patch, Charlie Moore's crew trawled for plankton samples and counted the numbers of plastic particles and planktonic organisms in them. Plastic outnumbered plankton six to one. In the winter of 2007–8 they repeated the exercise at the same spot. Moore reported that preliminary results showed "a significant increase in the ratio of plastic to plankton since 1999—as high as forty-nine parts plastic to one part plankton."

Moore also sent 270 of the plastic particles to a lab in Tasmania for chemical analysis. The good news is that 75 percent were polyethylene, 18 percent polypropylene, and 1.8 percent polystyrene—relatively stable plastics that aren't known to leach readily or disrupt the endocrine system. Only one particle was phthalate-packed PVC.

But this is just one small sample from one spot in the gyre. And the toxic and estrogenic properties of plastic particles don't end with their actual ingredients. Plastic, as Moore puts it and anyone who's used a plastic tub for washing dishes can affirm, "is a sponge for oily materials." Polyethylene pads are used to separate oil from water and soak it up when it spills. PCBs, DDT, and many other endocrine-disrupting compounds are oily. And so even the more stable plastics can become carriers for less benign chemicals.

In 2001, the organic geochemist Hideshige Takada and his colleagues at Tokyo University of Agriculture and Technology confirmed this hypothesis. They collected yellowed and discolored polypropylene nurdles from four beaches around Japan and measured them for PCBs, DDE (a breakdown product of DDT), and nonylphenol, a xenoestrogen used in cosmetics, cleaners, and other products. The levels of these in the nurdles correlated closely with levels in sediments and mussels at the various beaches. Takada proposed using pellets to inexpensively monitor pollution levels in water bodies.

His team then suspended new polypropylene pellets in highly contaminated Tokyo Bay for six days, periodically measuring for the same three contaminants. PCB and DDE, which were undetectable on the pellets before the trial, showed a "significant and steady increase" throughout it; nonylphenol levels were unchanged. Takada told *Science News* he thought plastic pellets might eventually accumulate concentrations a million times those in the surrounding water.

Oil spills surely loom in most minds as the ultimate marine disasters. That's especially true along Puget Sound, where we look anxiously up the coast toward Prince William Sound and imagine what havoc an *Exxon Valdez* might wreak if it cracked up in our constricted waters. On the other hand, people traditionally assumed that trash does little damage, other than looking ugly, unless it happens to be a bag or balloon or fishnet that can choke an animal.

And so I often shock people when I tell them that, as marine pollutants go, oil is relatively innocuous stuff. Certainly it looks awful as it coats the water, shore, and helpless seabirds, and it can cause grievous short-term damage. But oil dissipates and breaks down, becoming food for microorganisms. Vast quantities of oil were spilled in the Pacific and, especially, the Atlantic during World War II. Some beaches on the East Coast were coated several inches deep with the stuff. Not a trace of those spills remains today.

When we turn petroleum into plastic, however, we make it far more persistent and, I fear, more deadly. Animals take up the plastic nurdles and dust but cannot digest them, and slowly the toxins and hormone mimics leach out.

These effects are now starting to receive the attention they deserve through a slew of reports in the popular media on "the great garbage patch" (they typically describe the Eastern North Pacific Garbage Patch as though it were the only one in the world). Plastic is the new marquee villain, but even it is hardly the worst pollutant we feed into the sea. Mercury, lead, cadmium, and arsenic are elemental metals or metalloids that never break down, and they can have devastating toxic, neurological, and carcinogenic effects even in small doses. We are lacing the seas with them through industrial processes—from coal burning to pulp milling—and toxic trash. In the Gulf of Mexico, mercury-laced fluorescent tubes, perhaps from oil platforms, litter the beach at Matagorda Island. I found one that was crystal clear, its phosphor lining scrubbed clean away by the waves and its mercury gas long gone. In Oregon, the master beach scavenger Steve McLeod found one that still lit up when he plugged it in.

Electronic goods, especially cathode-ray computer monitors and televisions, are laced with lead, mercury, cadmium, and other heavy metals. A large monitor or picture tube may contain eight pounds of lead, embedded in its glass. Despite increasing restrictions on their disposal in many countries, monitors get tossed in landfills like everything else—and, like every other imported product, get washed off cargo ships. On January 28, 2000, seven containers holding about two thousand seventeen-inch monitors spilled into the North Pacific near the site of the 1990 Nike spill. (A seventh container washed up intact in Southeast Alaska, and a beachcomber salvaged and sold its contents.) The monitors' importer, Idaho-based Micron Technologies, was forthcoming with shipping details, and we could trace the progress of these baleful floating eyes. Between midsummer 2000 and 2001 I found one monitor at Grayland and received fifteen reports of wash-ups, often of multiple and in one case "many" monitors, from Oregon to British Columbia.

No matter how much flotsam you've scrutinized, you should never presume that you've seen everything the sea can throw up; the currents will always surprise you. Each year at the Ocean Shores Beachcombers' Fun Fair, I judge the Dash for Trash, a scavenger hunt that doubles as a beach cleanup and a forum for teaching people about seaborne litter and pollution. At the 2008 installment, several trash dashers returned with something strange: Thermos-sized aluminum canisters, label-less and scoured by the surf, with heavy red plastic caps. Some had unscrewed the lids, noticed white powdery residues inside, and sniffed to try to identify it. This was not a good idea; fortunately the residue was caked on and no one seemed to inhale any. I asked the pest-control guy who happened to have a booth next to mine if he'd like to examine a canister, and he recoiled in horror.

The canisters would have been cryptic, but I recognized them from a message the *Alert* had received from Wim Kruiswijk after similar canisters washed up in Holland in 2003. The white powder inside was aluminum phosphide, which when exposed to moisture produces phosphine gas, a potent insecticide and rodenticide. (This model's manufacturer calls it Gastoxin.) Grain shippers use the stuff to fumigate wheat and corn bound for Asia. They might uncap a couple hundred canisters, set them in the hold, and let the phosphine do its work on any roaches and rats.

This batch of canisters had actually begun washing up on Washington's shores the year before. After we got a warning out to the media, Coast

Guard, and beachcombing network, more reports poured in: twenty canisters found near Ozette, more than fifty at Copalis. State environmental officials figured they were tossed from a ship sailing out of Vancouver, BC, because the U.S. distributor of aluminum phosphide takes back the canisters it sells for reuse.

Alarming as this pest-bomb jetsam might seem, there was nothing illegal about tossing it in the sea, outside territorial waters. International law only forbids dumping plastic and certain specified pollutants—and aluminum phosphide isn't on the list. State officials saw no prospect of actually tracing and, perhaps, shaming the shipper responsible, but I hoped that the dates scrawled in pencil on some of the canister bottoms might provide a clue.

These weren't the first toxic canisters on local shores; beachcombers have since told me about others that washed up in 2000. I expect they won't be the last.

Since the early 1950s and 1960s, when humankind began to worry about its impact on the sea, one contaminant after another has seized public attention: mercury in the heyday of "Minamata disease," DDT, oil, PCBs, and now plastics and ocean-acidifying carbon dioxide. The point is not to fixate on any one toxic insult; the ocean remembers them all. We've got to combat all of them—to consider the cumulative effect they have on life in the sea and around it, and greatly expand our efforts to limit that effect. We might start by requiring that shippers and vessel operators report anything they lose or jettison in the sea. As it stands, they aren't even obliged to report the contents of lost shipping containers. They need only report the loss of eight or more containers at a time, because of the hazard to shipping.

12. The Music of the Gyres

The music crept by me on the waters.

—William Shakespeare, *The Tempest*

The laws of the abyss—have they been broken?

—Dante Alighieri, *Purgatorio*

Unless the sea's surface is glassy smooth, as the North Pacific almost never is, plastic particles mix in the waves and become virtually invisible to the eye. You can fly above a garbage patch (and if you've flown to Hawaii you probably have) and never know it's there. Nevertheless, two years ago the climatologist Jerry Norton asked me a provocative question: Might the garbage floating in a patch alter the reflection and re-radiation of light enough for satellites to read the difference from space? If so, it might finally be possible to precisely measure the extent of the garbage patches, something no one has yet found a way to do.

And if floating plastic does change the sea surface's reflectance, how will this affect the planetary ledger of heat absorption and reflection that underlies global warming? Mosslike marine invertebrates called bryozoans grow on plastic and, over time, whiten it until it comes to resemble snow. If this bleached plastic acts like snow and ice, boosting reflection, it might reduce planetary heat absorption and moderate to some degree the greenhouse effect.

Whether or not the debris collected within the oceanic gyres has any significant reverse-greenhouse effect, another effect is much more certain: Global warming will transform the movements of at least some of the gyres. How it will affect the eight open-water gyres is hard to predict, but the effect on the three icebound Arctic gyres is pretty clear—and it will surely be dramatic.

In the past fifty years, about half the Arctic ice pack has melted. It's a good bet that the other half will disappear in the next half century, allowing commercial shipping across the Arctic. Surface currents that are currently impeded by ice will speed up by a factor of thirteen, from a mean speed of 0.6 miles per day to 8 miles per day. This will cause a corresponding increase in the speed of the three now-icebound Arctic gyres and a drastic alteration in the basic music of planet earth—not just its rhythms but its harmonics. To understand this prospective effect, we must consider a remarkable, previously unnoticed aspect of the gyres.

As beachcombers sent me more reports of long flotsam drifts, I began organizing them by gyre, tabulating the orbits of thousands of drifters around the eleven gyres that make up most of the world ocean. From these I derived an average orbital period for each gyre. My rule of thumb was to start believing the averages when I got a sample of around twenty.

As each orbit table built toward twenty data points, I assembled a master table of average orbital periods. It revealed a surprising pattern, described here for the first time in any publication: the oceanic gyres' orbital periods occur in multiples of two. For the eight open-water gyres, these orbits reflect their relative circumstances. The smallest, the Viking (North Atlantic Subarctic) Gyre, makes one revolution in about 1.7 years. Five similarly sized gyres—the Columbus and Navigator in the Atlantic, the Aleut in the far North Pacific, the Majid in the Indian Ocean, and the Antarctica-circling Penguin Gyre—average 3.3 years per orbit. The two vast Pacific gyres, the northern Turtle and southern Heyerdahl, have average orbits of 6.3 years. The three Arctic gyres, slowed by their icepacks, have the longest orbits, 13.0 years.

I then compared the much shorter periods of the lesser gyres, ring currents, and frontal eddies that orbit within or spin off from the oceanic gyres. The same pattern of doubling held for them. It takes ten months for flotsam to travel around the great garbage patch in the Eastern North Pacific—half the orbital period of the smallest oceanic gyre, the Viking. It takes five months to round the garbage patches in the next five oceanic gyres—most famously the Sargasso Sea, in the Columbus Gyre—and so on down to the brief frontal eddies that break off from the Gulf Stream, which spin around in half the time of large transitory currents such as the gulf's Loop Current Eddy.

Incredulous, I tried to find contrary evidence, to disprove this four-step series of—depending at which end you start—doublings or halvings. But all the data I added only confirmed it. And then I noticed that it represented not only a rhythmic but a *harmonic* pattern. Take any length of material that resonates or is used to make air resonate to make a sound—a tube in a chime, a string in a violin, a wooden flute. Cut it in half, or fret it midway, or open a hole halfway down its length. You'll have raised its pitch by one *octave,* the most basic and nearly universal interval in music. Sound waves coming from it will cycle at twice the speed they did before. In common parlance, notes an octave apart sound like "the same note"—C and C, the "dos" that bracket re, mi, fa, sol, la, ti.

Even untrained ears—even monkeys—can recognize an octave; some researchers believe the interval is hardwired in the mammalian brain. Perhaps it runs even deeper in nature. The eleven classes of eddies, loops, subgyres, and oceanic gyres—each group taking twice as long to orbit as the last—form a global instrument with a prodigious range, ten octaves. By comparison, the most accomplished human singers have ranges of just four to five octaves. A piano's range is a little over seven.

As it happens, the human ear can perceive sound across a range of ten octaves—from twenty to twenty thousand cycles per second. But these are very different octaves from those sounded by the gyres; they spin at pitches billions of times lower than human hearing can detect. If you could speed a recorder enough to compress seventeen centuries into one second, the gyres would become audible. Only the human heart and imagination can hear the gyres; our senses perceive merely the flotsam they deposit on the shore.

Nevertheless, I contrived an audible toy to assist that effort of imagination. I've always loved wind chimes, and some years back my brother gave me a set that thereafter hung, little noticed, on the front porch, ringing at apparently random pitches. I took them down and cut the tubes to regularly proportioned lengths—1:2:4:8, and so on. Now they rang in octaves—Gaia's song, the music of the gyres, speeded up a billion times.

What will happen to this song as the planet warms? For the eight open-water gyres, the prospective effects are complex and somewhat unpredictable. More heat and greater temperature differences between air and water generate higher winds; already we see this process playing out and causing higher waves in the North Atlantic and North Pacific. Higher winds mean faster currents, which should speed up the gyres—but the effect is not uniform around the globe.

The melting of the glaciers and polar icepacks is sending more freshwater surging into the northern gyres, a flood boosted by increased flows from Siberia's rivers. In 2005, a study led by Harry Bryden of Britain's National Oceanography Centre found a growing volume of water flowing southward along the Canary Current into the North Atlantic Subtropical (Columbus) Gyre. This indicated a corresponding decrease—30 percent since 1957—in the current that carries warm water northeast from the Gulf Stream to the Viking Gyre and Europe, raising fears of another ice age there.

By the same token, summer—the space between the spring and fall transitions, when the prevailing winds shift between north and south—has lengthened along the Northeast Pacific, growing from 4.8 months in 1900 to 6.9 months in 2000. This means the summer northerlies blow longer from the southeast corner of the Aleut Gyre toward the Pacific Northwest, allowing more cool water to escape from the Aleut into the Turtle Gyre and slightly cooling the Turtle Gyre. But this effect is secondary to the flood of fresh water escaping from the Arctic into the Turtle and Columbus gyres. This added water piling up around the edges of the gyres will speed them up. Their orbital periods—their memories—will grow shorter.

I wonder if there is a global conservation of vorticity—if, as the currents speed up in one region, they slow down elsewhere. Harry Bryden has found that currents are slowing in the subtropical latitudes. Certainly they will speed up dramatically in the Arctic as the ice cap melts and ceases to brake them. This melting will expose the Arctic Ocean to wind, further accelerating the gyres. As they speed up, they will transfer warm water even faster to the cold latitudes, causing more melting and warming—a feedback loop.

Freed of their frozen burden, the three Arctic gyres will come to spin at the same average speed as their temperate counterparts, seven miles a day. It is windier there and so the orbital speed may be more like the antipodal cousin Penguin Gyre with a higher speed of ten miles per day. Their orbital periods will correspond roughly to their circumferences; the Melville and Storkerson gyres, similar in size to the Aleut and Viking, will likewise come to revolve once every 1.7 years in place of their current 13 years. The transpolar Polar Bear Gyre will accelerate from a 13-year to a 3.3-year orbit, matching the Columbus Gyre. The 13-year period that now governs the Arctic currents will be entirely lost.

If the eleven oceanic gyres are notes in a harmonic series, then that series will lose its fundamental tone, the lowest note upon which all the others rest. The global chorus will lose its bass, leaving just the baritone to undergird the tenors, altos, and sopranos. Make such an excision from a human chorus, knock the lowest octave out of a singer's range, and it would be a travesty. What words can describe such violence toward the earth herself? The gyres' music foretells a very different future from the gentle ride we have long enjoyed on their global carousel.

Forty years ago, Cliff Barnes taught me to cherish the water bodies of Puget Sound. Later, through research and practical oceanography, my horizons expanded, from Dabob Bay to the Strait of Juan de Fuca, the North Pacific, the North Atlantic, and finally, via the beachcombers' network, the entire floating world. As my understanding broadened, my love for the forms water takes, the movements it makes, and the treasures it transports grew ever deeper.

We will only save what we love, goes the classic conservationist syllogism. And we can only love what we know. Knowledge is power—the power to mend the world. This logic in some form or other drives and consoles the legions of scientists, naturalists, activists, and teachers who struggle, often in arduous conditions for meager compensation, to uncover and share the secrets of a natural world they see being assaulted and diminished daily.

Knowledge may cut to the core, but it is a double-edged blade. Our capacity to acquire it—the deuced cleverness of the human ape—also equips us to destroy what we should know and cherish. We have popped open the back of the magnificent watch that is the planetary ocean and begun to decipher the finely tuned workings of its intricate gears and springs. But we can destroy these exquisite works with a clumsy swipe of the hand, if we do not control ourselves.

As with watchworks, so with a sea or planet; all it takes is self-control. Moderate our appetites, cease trying to fill our lives with every new plastic-fantastic item, demand that the products we do need be responsibly produced and disposed of, and we will not fill the oceans and poison their denizens with junk.

Unfortunately, the precedents for the collective exercise of self-control by *Homo sapiens* aren't encouraging. How do we consider our prospects without despairing? If I look only at my generation and the other adult generations on this planet, raised in an ethic of blithe, unbounded consumption, I cannot avoid despair. But when I look at the children now approaching adulthood who will inherit the garbage patches and all the other messes we have made, I draw hope. They are growing up aware, at least dimly, of the consequences of our heedless actions, and they will demand solutions.

The Nike shoes and ducks and other bath toys have played some part in this. After hundreds of reports in children's magazines, books, and television shows, I'll bet virtually every youngster in North America and millions elsewhere have been exposed to the stories of the funny floaters—and hence to the concepts of oceanic gyres and garbage patches on a planetary scale. One after another, they walk up to me at beachcombing fairs and tell me they did their science project on the sneakers or tub toys.

Curt on his front porch, in a chair made from an old-growth cedar log found floating on Ross Lake, with some memorable flotsam: a Japanese fishing float, a Tommy Pickles Rugrat doll's head, a Nike cross-trainer, and tub toys lost in container spills.

We adults tend to fetishize flotsam curiosities, to make icons or mascots or collectibles out of them and forget the processes that produced them. But kids, who are supposed to be charmed by things like cute little duckies, see right through this. Time and again they tell me, "Gee, Dr. Duck, aren't they just trash?"

Talking to children renews my determination to decipher the secrets of the sea. This in turn takes me back to my own childhood—to the mini-gyres I created in my Flexible Flyer wagon, to the prophetic ducklings Flotsam and Jetsam, to the lessons my father and mother taught me. I only discovered the floating world because I listened to my mother when she asked me why all those sneakers were washing up along the Northwest coast. And we will only survive upon this water-blessed planet if we listen to our original mother, the great ocean, and the song she sings to us in the music of the gyres.

APPENDIX A

Urban Legends of the Sea

Over the years, I have come across untold strange, sensational, and moving accounts of ocean drifts. Many other authors have taken these tales at face value and offered them up as fact. I've tried to track down the actual facts of each case, weigh them against what I know about oceanic phenomena, and rate them from one to ten on a scale of probability. A rating of one indicates an assuredly false legend, five an account I give a fifty-fifty chance of being true, and ten an account that is confirmed beyond doubt. I have based this book on accounts I value at ten, with exceptions duly noted. Further investigation may raise the others to higher values, but thus far, after twenty years' work, this is where I stand on six of the most celebrated and persistent tales of the ocean's currents:

Theophrastus's Bottles. For years, I cherished Gardner Soule's *Men Who Dared the Sea*, which recounted how Aristotle's protégé Theophrastus released bottles in the Strait of Gibraltar to see if a current flowed from the Atlantic into the Mediterranean. Soule included footnotes for many parts of his book, but none for his chapter on Theophrastus. Still I kept investigating; this experiment—which would have made Theophrastus the first by about two millennia to make scientific use of drifters—seemed in line with the way he conducted research. And the 1911 *Encyclopedia Britannica* gave the same account. To confirm it, I contacted a group of scholars at Rutgers University, who were assembling all Theophrastus's known writings. They graciously investigated but could not verify that he released drifters. And so I rate this beguiling tale a mere four.

The Royal Uncorker. The tale is often told that in 1560 Queen Elizabeth I of England charged an official with opening any message bottles that drifted on Albion's shore, in case they might have come from her spies on the continent. Anyone else uncorking a message was subject to death. I traced this account to a story by Victor Hugo, which contained many details that seemed somewhat plausible. Try as I might I have not been able to trace it further, and I give it just a three.

The Ghost Ship *Octavius*. In 1775, the oft-told tale goes, the whaling ship *Herald* was becalmed off Greenland's east coast when a mysterious empty ship drifted into view. When it did not reply to signals, the *Herald*'s captain dispatched a boarding party. As the scouts approached, they read the ship's faded name: *Octavius*. Once aboard, they found the crew below deck, frozen solid. The captain sat frozen at his table, pen in hand and log before him, with the bodies of his wife and child beside him. The boarders hurried from the ship, taking only the captain's log. The last entry in it was thirteen years old. The *Octavius* had left England for the Orient in 1761 and arrived the next year. The captain then risked a shortcut no one had succeeded in crossing: the fabled Northwest Passage. He and his crew were the first to sail the passage—posthumously, victims of the Arctic ice. After its encounter with the *Herald*, the *Octavius* was never seen again.

From an oceanographic view, this tale is surprisingly plausible. Thirteen years, the supposed drift of the derelict *Octavius*, is just one revolution of the great, ice-clogged Storkerson Gyre north of Alaska and Canada. The ship could conceivably have completed half a circuit, then gotten picked up by the great Polar Bear Gyre that circles the Arctic Ocean and carried east to Greenland. Otherwise, however, the tale seems wildly fanciful, and serious authorities dismiss it as fiction. I accord it a two, at the bottom of the probability scale.

The Phantom Lifebelt. The light cruiser *Sydney*, one of Australia's most famous warships, was sunk in battle off Western Australia in 1941, with all hands lost. While investigating drift patterns in the Southern Hemisphere, I came across some oceanographers who had explored the *Sydney*'s wreckage. They spoke of a lifebelt that had supposedly drifted all the way to France—one of the longest recorded drifts in history. The time frame seemed reasonable, so I investigated further. I tracked the lifebelt to French coast guardsman Henri Pouelet, who found it washed up at Saint Gilles-sur-Vie in January 1951. Pouelet's son believed the story to be true, but the lifebelt itself was nowhere to be found. Apparently, the French government had confiscated it from his father, and that was the last clue I could find. I rate this one a seven—almost, but not quite confirmed. Many lifebelts must have washed ashore in the war, and there's no way to be sure this one came from the *Sydney*.

The $6 Million Bottle. Radio and TV hosts often ask, "What's the most valuable prize ever found in a bottle?" The answer is certain—if the story that my clipping service (my mother) first spotted in the *Environmental Almanac* is true. It begins with the polygamous sewing-machine magnate Isaac Merritt Singer, who fathered five broods and managed to keep them apart until 1860, when two of his mistresses spied each other from carriages traveling along New York's Fifth Avenue. When he died, Singer left substantial fortunes to all his fourteen illegitimate children. The youngest, Daisy Alexander (née Mary McGonigal), seems to have inherited a bit of her father's tendency toward eccentricity. She grew up in San Francisco, married into the English aristocracy, and moved to London, where she took to throwing bottled messages into the Thames River. When she died in 1939, at the age of eighty-one, her will could not be found—not surprisingly, since it was floating around the Arctic.

By 1949, Alexander's estate had swollen to $12 million. That year, Jack Wurm, the penniless owner of a bankrupt restaurant, happened to be walking along San Francisco Bay when he spied a bottle with a scrap of paper in it. He broke it open and read, "To avoid all confusion, I leave my entire estate to the lucky person who finds this bottle and to my attorney, Barry Cohen, share and share alike. Daisy Alexander. June 20, 1937." The bottle had drifted for twelve years and ten thousand miles, down the Thames, across the North Sea, around Scandinavia, Russia and Siberia, through the Bering Strait and Sea, and down the Pacific Ocean to San Francisco, where Daisy Alexander grew up. Wurm claimed his half, but because Alexander had failed to notarize her will, the Singer Company disputed it. Wurm finally prevailed and allegedly received $6 million in cash and Singer Company stock.

I became fascinated with this bottle's drift, which would have skirted four gyres: two in the Arctic and the Aleut and Turtle gyres in the Pacific. More remarkable yet, I have identified six other drifters that traveled substantially the same ten-thousand-nautical-mile path. I even talked with a finder of one of the other bottles. The average time for these drifts was close to the twelve years Daisy's bottle supposedly floated. Wherever the story came from, it included accurate, very rare drift data! But I could find no newspaper or court records or other confirmation of the case. Based on the six similar drifts I could document, I give this one a five—an even chance of being true.

"Where there's a way, there's a will," noted Jim Ingraham.

Pang's Wheel. One evening in 1993, after I gave a talk in Poulsbo, Washington, an elderly gentleman with a prodigious memory came up and said he remembered a *Ripley's Believe It Or Not!* comic back in the early 1930s that might interest me. As he recalled, when the pioneer aviator Clyde Pangborn made the first nonstop flight across the Pacific, he jettisoned his plane's wheels to reduce drag, and one then drifted from Japan to Washington. That was all I had to go on.

To confirm the tale I first had to locate the *Ripley's* installment, date unknown. After plowing for days through library microfilm, I finally found it. It was just as the old gent remembered, but I was skeptical; *Ripley's* provided no identifying details for the wheel. Tracing it would not be easy; many wheels must have gone adrift in the Pacific. How could I be sure the one found on the Washington coast was the same one Pangborn jettisoned?

After he died in 1958, Pangborn's papers wound up at Washington State University. I contacted the archivist, hoping to find out where the actual wheel was. Surely such a famous artifact had been preserved! It had, by the Firestone Company, which made the tire. Unfortunately, Firestone had since been sold to Bridgestone, and the wheel seems to have been discarded in the merger.

I tracked down Carl Cleveland, a Pangborn pal who had written a book about him and who happened to be living in Seattle. Cleveland still had many photos of Pang, as he called him, and let me take what I wanted. I chose those showing details of his plane, *Miss Veedol*, being prepared for takeoff from the beach at Sabishiro, Japan. The photos showed details of the wheel-release apparatus. One revealed the serial numbers on the tires.

Then I learned that a legal fracas had ensued over the recovered wheel, and the court records were here in Seattle. Sure enough, they included the tire's serial number, matching one in the photo. And the drift shown in *Ripley's* agreed with many other drifters along the same pathway in both time and location. Believe it, indeed; someone at *Ripley's* did his homework. And notch up this entirely persuasive account as a ten.

APPENDIX B

A Million Drifting Messages

Where message bottles were launched and found, how many were launched, return rates, and rewards offered for their return.

This table compiles known data for thirty-two of the thousands of drifter launches conducted worldwide in the past century and a half—a total of 999,677 message bottles. About 31 percent—305,951—were launched by evangelist "gospel bombers." This tabulation does not include thousands of drifting buoys tracked by satellite or many thousands of messages released by schoolchildren in Japan and by the navies of Japan, the United States, Britain, and Germany.

These launches are ordered by return rate, from unknown to lowest to highest rates.

Area of launch and/or launcher	Rate of return (number launched)	Reward for reply	Location launched
Mrs. Gause	unknown (500)	unknown	Plant City, Florida
Walter E. Bindt	unknown (801)	unknown	San Francisco
Ethel Tinkham	unknown (3,950)	unknown	Portland, Oregon
Printed Page Evangel Society	unknown (20,500)	pamphlets	Belfast, Ireland
Captain Walter Bindt	unknown (many thousands)	pamphlet	worldwide
Nigel Wace	1 percent (1,400)	none	Drake Passage, Antarctica
South Africa	1.6 percent (50,000)	none	Cape of Good Hope
Australia (John Bye)	2.2 percent (91,000)	unknown	Southern Ocean off Australia

Site of launch and/or launcher	Rate of return (number launched)	Reward for reply	Location found
Kraig Voigt Rice	4 percent (92,400)	bibles, clothing	California to Philippines
Schwartzlose (Scripps Inst.)	4 percent (201,034)	none	coastal California
Ocean Station Papa (1956–59)	5 percent (33,869)	unknown	northeast Pacific
George Tinker	5.1 percent (1,400)	none	60 miles off Oregon
Alan Schwartz	6 percent (233)	$1 (new)	Southern California
J. B. Matthews	7 percent (2,070)	unknown	Arctic coastal Alaska
George Phillips	7.5 percent (40,000)	spiritual pamphlets	worldwide
Merseyside Bottle Evangelists	8.5 percent (65,000)	spiritual aid	worldwide
Drift casks	9.4 percent (32)	none	Arctic Ocean
Everett Bachelder	10 percent (60,000)	spiritual aid	Bering Sea
Dean Bumpus (Woods Hole)	10 percent (165,566)	$0.50	U.S. East Coast
Seamen's Church Institute	12 percent (5,000)	paintings	North Atlantic
Oregon State University	14 percent (21,615)	none	coastal Oregon
Gulf of Mexico scientists	15 percent (85,000)	minimal	Gulf of Mexico & Caribbean
Judge Boyd C. Baird	16 percent (68)	$1 for postage	mid-ocean, worldwide
Deborah Diemand	18 percent (8,600)	none	Newfoundland/ Labrador to Europe
Louanne Peck	20 percent (30)	unknown	mid-North Pacific
Jewel T. Pierce	20.3 percent (27,800)	spiritual aid	Coosa River, Alabama
Avalon's Bicentennial	24 percent (200)	coin set	coastal California
U.S. Geological Survey	32 percent (2,780)	unknown	San Francisco Bay
Long Island Sound	38 percent (154)	unknown	Long Island Sound
Drifting buoys	44.4 percent (unknown)	reward label	Northern Gulf of Mexico
North Sea drifters	50 percent (unknown)	unknown	North Sea, Hakan Westerberg
Puget Sound drift cards	50 percent (17,000)	none	Puget Sound

APPENDIX C

The Oceanic Gyres

Average orbital periods, dimensions, and orbital speeds of the eleven main oceanic gyres.

Gyre	Orbital period (years)	Length and width (nautical miles)	Circumference (nautical miles)	Orbital speed (miles per day)
Polar Bear (Transpolar)	13.0	500 x 2,400	6,000 (Ice-covered portion: 4,800 Open water: 1,200)	0.6 7.0
Melville	14.1	200 x 1,200	3,000	0.6
Storkerson (Beaufort)	12.1	500 x 1,000	2,500	0.6
Turtle	6.5	1,600 x 5,000	14,000	5.9
Heyerdahl	6.5	2,200 x 5,100	14,750	6.2
Penguin (Antarctic)	3.7	3,600 x 3,600	12,000	8.9
Majid	3.7	1,200 x 3,600	10,000	7.4
Navigator	3.0	2,100 x 2,400	9,500	8.7
Columbus	3.3	1,200 x 3,000	8,000	6.6
Aleut	3.0	800 x 3,500	6,800	6.2
Viking	1.7	1,000 x 1,200	4,500	7.3
Total	**70.6**		**91,050**	

APPENDIX D

Ocean Memory

Based on eleven long-term drifters that orbited six of the eleven planetary gyres.

These drifters have been ordered according to gyre memory—the fraction of drifters remaining in a gyre after one orbit. Imagine 1,000 toy ducks released in the Aleut Gyre, which has an orbital period of three years. The flock will circle the gyre every three years. Averaging the two estimates of the Aleut Gyre's memory (0.40, 0.44) indicates that after the first orbit about 420 of the 1,000 ducks will remain in the gyre. After the second orbit, 42 percent of these 420 ducks or 176 ducks will remain. And so on till the last duck drops out.

The ice-bound Arctic gyres, whose orbits take much longer, appear to have much shorter memories; nearly 90 percent of drifters are lost from the Storkerson and Melville gyres with each thirteen-year orbit. Outside the Arctic, however, gyre memory varies twofold, from 0.34 to 0.80, and averages 0.50, based on ten estimates in this table. This means that half the drifters remain in a gyre after each orbit, a ratio recalling radioactive decay. Thus, for the eight open-water gyres, half of a group of drifters remain after the first orbit, a quarter after the second orbit, an eighth after the third orbit, one in a thousand after ten orbits, and so on till the last drifter leaves the gyre.

Statistically speaking, the concept of gyre half-life accounts for an average of 94 percent of the variability shown by the timeline of each long-term drifter. This means there is a 1 percent chance that what appears to be gyre memory is merely random. That all drifters fit the pattern to such a high degree means that chance can be ruled out. No doubt memory varies between the gyres, but we don't yet have data on enough long-term drifters to assess these variations.

Gyre	Memory (fraction remaining after one orbit)	Fraction remaining after 10 years	Number of orbits made by oldest drifter	Type of drifter	Longest drift (years)	Number recovered
Storkerson, Melville	0.11	0.18	1.5	drift cards	20	36
Turtle	0.34	0.19	4.5	message bottles	29	103
Columbus	0.38	0.053	8.2	lobster tags	27	119
Aleut	0.40	0.047	4.8	tub toys	16	120
Aleut	0.44	0.065	5.2	bottles	17	97
Penguin	0.45	0.14	4.9	cards	18	199
Turtle	0.48	0.32	6.5	clay jars	42	61
Columbus	0.52	0.14	9.7	cards	32	183
Columbus	0.55	0.16	12.7	bottles	42	152
Turtle	0.61	0.47	4.6	OSCURS drifters	30	1,120 (30-year tracks)
Turtle	0.80	0.71	8.6	glass balls	56	tens of millions

APPENDIX E

Harmonics of the Gyres

Gyres and other orbiting currents classified by tone—length of orbital period—from longest to shortest.

Octave, Tone	Average Orbital Period	Type of Vortex: Gyre, Ring, Eddy	Examples and Effects
1) Fundamental tone	13.0 years	Melville, Storkerson, Polar Bear gyres	Longest drift times of all. Global warming and ice melt will eliminate this fundamental tone.
2) Half tone	6.5 years	Turtle, Heyerdahl gyres	Live mines, derelict junks, and fifty-year-old flotsam have circled the North Pacific's Turtle Gyre. Thor Heyerdahl rafted around its southern counterpart.
3) Quarter tone	3.3 years	Columbus, Navigator, Majid, Penguin gyres	Columbus Gyre helped Columbus reach America.
4) 8th tone	1.7 years	Viking Gyre	Vikings reached America along its northern arc.
5) 16th tone	10 months	Great Eastern North Pacific Garbage Patch	OSCURS shows that this patch stores flotsam for 30-plus years.

Octave, Tone	Average Orbital Period	Type of Vortex: Gyre, Ring, Eddy	Examples and Effects
6) 32nd tone	5.0 months	Various garbage patches, possibly fixed subgyres	Sargasso Sea is best-known example.
7) 64th tone	2.5 months	Fixed subgyres such as Gulf of Alaska and other bights	Pumice from Katmai eruption found on seafloor of Gulf of Alaska.
8) 128th tone	1.3 months	Subgyre around Iceland (400-mile diameter at 36 miles per day)	Drifting heirlooms aided Vikings in selecting home sites.
9) 256th tone	19.1 days	Large Gulf Stream rings (300-mile diameter at 48 miles per day)	Rings can circulate for years in the Sargasso Sea.
10) 512th tone	9.6 days	Large transitory eddy: Gulf Stream ring, loop current eddy	"Freedom floaters" fleeing Cuba have been fatally diverted on loop current eddies.
11) 1,024th tone	4.8 days	Gulf Stream frontal eddy	Smaller eddies pinching off Gulf Stream.

Acknowledgments

More people helped us in the course of this long journey than can be credited in a short book. John Roy (Jack) Beck and Eugene E. Collias at the University of Washington and Ronald E. Westley at the Washington State Department of Fisheries greatly aided my snark hunt. At Mobil Research and Development Corporation, Jack Hubbard took me under his wing and initiated me into the world of offshore oil-structure design, and Cornelius W. Langley provided elegant mathematics on the spread of oil on the sea that happened to prove applicable to snarks.

Thousands of beachcombers worldwide have subscribed to the *Beachcombers' Alert!* and supplied vital flotsam data, including Nick Darke and Stella Turk (Cornwall), Louis E. Hitchcock (Wake Island), Judie Clee (Bermuda), Cathy Yow (Jamaica Beach, Texas), Margie Mitchell (Cocoa Beach), Scott Walker (Ketchikan), Janet Etzkorn (British Columbia, Canada), Kay Gibson (Camden, Maine), Doris Hannigen (Olympia), John Anderson (Forks, Washington), Neil and Betty Carey (Queen Charlotte Islands), Gene Woodwick (Ocean Shores, Washington), Michael Armstrong (Homer, Alaska), Dean Orbison (Sitka), Wim Kruiswijk, Irene Maas, and Henk Noorlander (the Netherlands), Mark Michael (Rota Island), Martin Huebeck (Shetland), and Judy d'Albert (Harbor Day School in Corona del Mar). Carol Wickenhiser-Schaudt supplied illustrations of flotsam.

Thanks also to the many organizers of the more than fifty beachcomber fairs I have attended: Sea-Bean Symposium in Cocoa Beach, Florida; the Beachcombers' Fun Fair in Ocean Shores, Washington; the Driftwood Show in Grayland, Washington; and the Beachcombers' Fair in Sitka, Alaska, where vast numbers of clues to the floating world surfaced.

Our expeditions to some of the trashiest beaches on earth would not have been possible without friends who provided hospitality and base camps: for my Mexican adventures, Christopher Boykin and Marcia Bales at the Mayan

Beach Garden in Costa Maya; for Matagorda Island, Sam and Mike Barnett in Port O'Connor, Texas; on our Lanai adventure, Captain Rick Rogers; in the search for a Manila galleon lost off Baja California, Edward P. Von der Porten of San Francisco; skipper Larry T. Calvin, who brought us to the beaches around Sitka aboard the *Morning Mist*; Captain Charles Moore of the *Alguita*, who brought Jim Ingraham and me to sample plastic in the Northwest Hawaiian Islands; and, of course, Noni and Ron Sanford on the Big Island.

Several stalwarts keep the *Beachcombers' Alert!* going out to the world: James R. White, managing editor; David Byng Ingraham, photo editor; Kari Sauers and Tim Crone, webmasters; Dave McCroskey, treasurer; Sally A. Mussetter, who proofreads it; and Jan White, who manages its mailing list. Others have helped me prepare scientific articles, including Jeffrey M. Cox, Carol Coomes, Brent Johnston, David W. Thomson, Timothy J. Crone, and Jonathan M. Helseth at Evans-Hamilton; Glen N. Williams at Texas A&M University; Charles D. Boatman, a consultant in Seattle; and Glenn A. Cannon at NOAA's Pacific Marine Environmental Laboratory in Seattle.

Ella Phillips shared her late husband George Phillips's letters and photos. Roy Overstreet and Debbie Payton opened NOAA's historical files on, respectively, Arctic drift cards and cards released to track hazardous substances. John A. T. Bye at the Flinders University of South Australia supplied many Southern Hemisphere drifters. Nigel M. Wace of National University at Canberra shared data on his epic drifters around the Penguin and Heyerdahl gyres. Carl M. Cleveland provided photographs of his friend Clyde Pangborn. Katherine Plummer, author of *The Shogun's Reluctant Ambassadors*, shared her extensive knowledge of Ranald MacDonald and the drifts of Japanese vessels across the North Pacific. Lynn De Witt of NOAA's Pacific Fisheries Enviromental Group supplied navy pressure fields, OSCUR's bedrock data, to Jim Ingraham for thirty years.

Many members of the media have helped get out the word about sneakers, tub boys, Lego blocks, and other telltale flotsam washing ashore. In Seattle, Steve Scher and Katy Sewall, host and producer of KUOW-FM's daily *Weekday* and occasional "Flotsam Hour," did more than they know to set this effort in motion. Susie Ebbesmeyer and Maria Lucia Hansen assisted us with grace and wisdom at every stage, from combing beaches to brushing up rough drafts.

We owe special debts to our patient editor, Elisabeth Dyssegaard, who understood this project from the start and made it better at every stage, and our indefatigable agent, Elizabeth Wales, who went above and beyond the call of her profession to bring the floating world to the printed page.

Illustration Credits

Page 3: Ebbesmeyer Family photo
Page 5: Pi Kappa Tau photo
Page 11: Curtis Ebbesmeyer PhD dissertation
Page 13: Curtis Ebbesmeyer illustration
Page 23: Barnes Family photo
Page 30: Ebbesmeyer Family photo
Page 33: Evans-Hamilton photo
Page 41: Paul Ebbesmeyer illustration, published in the *Journal of Physical Oceanography*, March 1986
Page 47: Bruce R. Johnson photo
Page 54: Paul Ebbesmeyer illustration
Page 57: Naval Historical Center photo; G. W. Melville drawing
Page 59: Nancy Hines photo
Page 66: Nancy Hines photo
Page 71: Jim White photo
Page 76: Curtis Ebbesmeyer illustration, Jim Ingraham simulation, published in *EOS* August 25, 1992
Page 80: Dave Ingraham photo
Page 82: Curtis Ebbesmeyer illustration, Jim Ingraham simulation
Page 88: Bruce R. Johnson photo
Page 99: Curtis Ebbesmeyer illustration
Page 102: Dorothy Lang Orbison photo (left); Judith Selby-Lane photo (right)
Page 112: Museum of Industry and History, Seattle
Page 114: Jim Ingraham photo
Page 121: Kell Ingraham photo
Page 125: Cliff Barnes photo
Page 143: Jim Ingraham illustrations
Page 148: Museum of Industry and History, Seattle
Page 150: Jim Anselmino photos
Page 154: *Chinook Observer* photo
Page 158: Dave Ingraham illustration
Page 168: Dave Ingraham illustration
Page 170: Dave Ingraham illustration
Page 178: Leah Keshet photo
Page 193: Dave Ingraham illustration
Page 200: Dave Ingraham photo
Page 204: Joel Paschal photo
Page 211: Susan Middleton photo
Page 217: Eric Scigliano photo
Page 228: Dave Ingraham photo

Glossary

Alert Network: Thousands of beachcombers worldwide who report washed-up flotsam. Their findings appear in the quarterly newsletter *Beachcombers' Alert*.

Banksman: Shetland term for beachcomber. (Source: birder Martin Heubeck.)

Beach bum: How others often see those who beachcomb.

Beaner, bean head: What collectors of sea beans call themselves and each other.

Beachcomber: One who gathers, collects, studies, or examines objects on a beach. Also banksman (Shetland), wrecker (Cornwall), jutter (Dutch), raquero (Spanish), and comber (United States).

Bottle paper: Message in a bottle (historic).

Bryozoans: Mosslike marine invertebrates that grow on plastic flotsam and coat it white.

Clast: Pumice fragment transported by ocean currents.

Comber: Beachcomber (United States).

Confetti: Small chips of plastic, an intermediate stage as plastic flotsam crumbles into dust.

Container unit: The volume of a standard 8-by-8-by-40-foot shipping container (67.5 cubic meters; 88.2 cubic yards). Synonym: **FEU** (forty-foot-equivalent unit).

Convergence: A place where opposing currents meet. For example, two currents meeting head-on often form rip lines of flotsam. The flotsam riding the current of denser water stays on the surface as the current dives below the other current of lighter density. Also, where opposing air masses meet.

Conveyor belt: The current system interconnecting the eleven planetary gyres. It is the system whereby the gyres hand off flotsam between one another and so on around the planetary ocean.

Current paper: Message in a bottle (historical).

Derelict: A vessel abandoned at sea.

Density, water: A cubic yard of water—barely enough to surround two people seated with their legs crossed—weighs nearly a ton. A cubic foot of seawater at the surface weighs on average 63.9 pounds, and a cubic yard contains 27 cubic feet, or 1,726 pounds.

Drift line: The path followed by a drifter. Often a line of flotsam on the ocean, demarcating a drift.

Driftographer: One who studies all things afloat on the sea.

Dust: In oceanography, flotsam.

Eddy: A circular current, such as a whirlpool, in the ocean or another water body.

Extinction level event (ELE): A catastrophe so severe as to cause mass extinctions on a planetary scale, such as the volcanic eruption that formed Lake Toba and may have caused the last ice age.

Eulerian: Fixed in location, as a reference frame.

FEU (forty-foot-equivalent unit): The volume of a standard 8-by-8-by-40-foot shipping container (67.5 cubic meters; 88.2 cubic yards). Synonym: container unit.

Flotsam: Something floating lost accidentally at sea.

Flotsamology: The study of flotsam (coined by Wendy Ebbesmeyer on July 8, 2003). **Flotsamologist.**

Flotsametrics: The quantitative study of flotsam (coined by Jim Ingraham on February 4, 2001). See also **sinkametrics** (coined in rejoinder by Steve Cummings). **Flotsametrician.**

Foamer: Railroad enthusiast and, by extension, beachcomber (slang).

Garbage patch: Region inside a gyre where drifting objects collect. Major garbage patches are often under semipermanent cells of high sea-level atmospheric pressure. (Term coined by Curtis Ebbesmeyer in the 1990s.)

Geobean, geoseed: Long-range drifting seed.

Geodrifter: Long-range drifter that can cross an ocean.

Geoflotsam: Long-range flotsam.

Grand Tour: Long, continuous drift pathway, circling the earth, with flotsam passed from one gyre to the next passing through all the oceans.

Gyre: Continental-scale closed loop of water around which flotsam drifts. The gyres vary eightfold in diameter from one thousand nautical miles along the major axis of the Storkerson Gyre in the Arctic Ocean, to five thousand nautical miles across the Heyerdahl and Turtle Gyres in the South Pacific. We have chosen to assign the eleven largest gyres new names that are more easily remembered and more reflective of their distinctive drifter characters and histories.

The Eleven Oceanic Gyres:

Ocean	*Name assigned in this book*	*Traditional oceanographic designation*
Arctic	Storkerson Gyre	Beaufort Gyre
Arctic	Melville Gyre	None
Arctic	Polar Bear Gyre	None
North Pacific	Turtle Gyre	North Pacific Subtropical Gyre
North Pacific	Aleut Gyre	Pacific Subarctic Gyre
South Pacific	Heyerdahl Gyre	South Pacific Subtropical Gyre
North Atlantic	Columbus Gyre	North Atlantic Subtropical Gyre
North Atlantic	Viking Gyre	Atlantic Subarctic Gyre
South Atlantic	Navigator Gyre	South Atlantic Subtropical Gyre
Southern	Penguin Gyre	Antarctic Circumpolar Gyre
Indian	Majid Gyre	Indian Subtropical Gyre

Gyre memory: The share of drifters retained in a gyre after each orbit. A memory of 0.5 (the global average) indicates a gyre retains 50 percent of its drifters through each orbit while the other half washes up on shore, escapes to other gyres, or sinks. Half remain after the first orbit, a quarter after the second orbit, an eighth after the third orbit, and so on until the last drifter strands.

Hadley cell: A circulation pattern that dominates the tropical atmosphere, with rising motion near the equator and air flow toward the equator near the surface and toward the poles ten to fifteen kilometers above the surface.

Hyperspeed: In flotsamology, more than twenty-five nautical miles per day unless otherwise stated, the speed of the fastest drifters traveling the Grand Tour.

***Hyôryû*:** In Japanese, a marine mishap in which a vessel (*hyôryû-sen*) loses control and drifts on the high sea.

***Hyôryû-min*:** "Drifting people," the hapless passengers aboard a *hyôryû-sen*.

ICDM, ICFM: Intercontinental drifting/floating mine.

International Geophysical Year (IGY): An international scientific effort lasting from July 1, 1957, to December 31, 1958, and encompassing eleven earth sciences, including oceanography.

Jetsam: Floating objects jettisoned deliberately, typically to save a threatened vessel.

Jutter: Beachcomber (Dutch).

Knot: In navigation, nautical mile per hour, equaling 1.15 statute miles per hour.

Lagrangian (describing a spatial reference frame): Attached to a drifting object rather than a fixed site, mobile.

Light stick: A plastic tube containing fluids that glow when allowed to mix, used to attract fish.

Message in a bottle: A missive or document set adrift in a sealed bottle.

Miles: Nautical miles, unless specified as statute miles.

MIB: Message in a bottle (coined by Curtis Ebbesmeyer).

Nautical mile: In distance measurements, 1.15 statute miles, 1,852 kilometers.

Neutral buoyancy: When the density of a drifter equals that of surrounding water.

NIV: Bible, New International Version.

Nurdle: Pellet of preproduction plastic (coined in the 1950s by Southern California lifeguards).

Octave: An interval in a series of halved or doubled quantities. For example, gyre orbital periods of 1.6, 3.3, 6.7, and 13.4 years are one octave apart.

Orbit: The outer circumference of a gyre.

Orbital period: The time required for a floating object or water slab to drift once around a gyre, reflecting the cumulative effects of currents, winds, and waves.

OSCURS: Ocean Surface Current Simulator, a computer program developed by W. J. (Jim) Ingraham, Jr., to reconstruct and predict the movements of ocean currents and their flotsam.

Raquero: Beachcomber (Spanish).

Rip lines: Lines at the sea surface where currents meet.

Sea bean: One of several hundred species of seed or fruit that float on seawater as part of their propagative strategies. Twenty or so species can float for years, long enough to cross an ocean.

Snark: Water slab observed in Dabob Bay, Washington, by Curtis Ebbesmeyer.

TEU (twenty-foot-equivalent unit): The volume of an 8-by-8-by-20-foot shipping container (1,140 cubic feet; 32.3 cubic meters; 42.2 cubic yards). Though most containers are now twice this size, TEU remains the standard measure of container volume.

Water chaser: An oceanographer who tracks blobs or slabs of water—often eddies—using drifters and measurements of temperature, salinity, and dissolved oxygen.

Water slab: A chunk of water distinguished from surrounding water by such properties as temperature, salinity, density, dissolved oxygen, color, and smell.

Windage: The drag exerted on a drifter by wind, distinct from the drag exerted by underlying currents.

Wrecker: Cornish term for one who is interested in all things on the beach, beachcomber. Contrary to popular misconception, it did not describe people who lured boats onto the rocks; Cornish shore dwellers did not carry on this practice. (Described in the film *The Wrecking Season* by Nick and Jane Darke.)

Xenoestrogens: External substances, including several marine pollutants, that bond to receptors for the human estrogen estradiol, mimicking the action of this hormone.

Further Reading

A thematic, annotated, selected bibliography.

Ashes

As I tossed Akira Okubo's ashes to the winds over Lake Diablo, I observed that some were as fine as the finest clay particles. According to table 105 in *The Oceans*, the oceanographers' bible by Sverdrup, Johnson, and Fleming, clay particles measuring 0.12 microns settle at the rate of one millimeter per day. Assuming that the surface waters circulating Turtle Gyre reach down to depths of one thousand meters, such particles would remain in this layer for 2,740 years, or approximately five hundred orbits of the Turtle Gyre. I am not aware that anyone has published the grain-size analysis of human ashes. For burial at sea, cremations should be accomplished so as to produce the finest possible ashes, thereby insuring the longest floating and settling times. Sverdrup, H. U., Martin W. Johnson, and Richard H. Fleming, 1942. *The Oceans: Their Physics, Chemistry, and General Biology.* Englewood Cliffs, New Jersey: Prentice Hall.

Azores Garbage Patch

Rudolf Scheltema analyzed 480 transatlantic message bottles collected in the Woods Hole Oceanographic Institution's archives. Scheltema, Rudolf S., 1966. "Evidence for Transatlantic Transport of Gastropod Larvae Belonging to the Genus *Cymatium*." *Deep-Sea Research* 13: 83–95.

Prince Albert I analyzed 227 bottles retrieved from the North Atlantic. H.S.H. Albert, Prince of Monaco, 1892. "A New Chart of the Currents of the North Atlantic." *Scottish Geographical Magazine* (October 1892): 528–31.

The colonial botanist Henry Guppy carefully sought out rare accounts and recorded the drifts of many bottles. Guppy, H. B., 1917. *Plants, Seeds, and Currents in the West Indies and Azores: The Results of Investigations Carried Out in Those Regions Between 1906 and 1914.* London: Williams and Norgate.

Bible

Goodrick and Kohlenberger enumerate 665 biblical references to water and 355 to the sea on pages 808–9 and 989–92 of their concordance. Goodrick, Edward W., and John R. Kohlenberger III, 1981. *The NIV Complete Concordance.* Regency Reference Library. Grand Rapids: Zondervan Publishing.

Buoys

From 1871 to 1884, thirty-one buoys marking shoals and channels tore loose from North American coastal waters and drifted around the Columbus Gyre. Johnson, A. B., 1884. "North Atlantic Currents." *Science* (October 31, 1884): 415–18.

Columbus, Christopher

See also Azores Garbage Patch, Henry the Navigator.

Biographies. *The Life of the Admiral Christopher Columbus by his Son Ferdinand.* Translated, annotated, and with a new introduction by Benjamin Keen. New Brunswick, New Jersey: Rutgers University Press.

Morison's is the essential modern biography. Morison, Samuel Eliot, 1942. *Admiral of the Ocean Sea, a life of Christopher Columbus.* Boston: Little, Brown.

Columbus's weather luck. Dr. Charles F. Brooks, an experienced meteorologist, worked with Morison, who resailed Columbus's voyage, to reconstruct Columbus's route through two storms. Brooks, Charles F., 1941. "Two winter storms encountered by Columbus in 1493 near the Azores." *Bulletin of the American Meteorological Society* 22 (October 8, 1941): 303–9.

Container Spills

Nike cross-trainer spill, 1990. My mother's questions about the washed-up sneakers, my inquiries about the *Hansa Carrier,* my fax to my old grad schoolmate Jim Ingraham, and his blind test of OSCURS predicting sneaker washups led to this paper. Ebbesmeyer, C. C., and W. J. Ingraham, 1992. "Shoe

Spill in the North Pacific." *EOS, Transactions of the American Geophysical Union* 73 (34): 361–65.

This textbook article reached 40 percent of college students studying beginning oceanography. Ebbesmeyer, C. C., 1994. "The great sneaker spill." In *An Introduction to the World's Oceans* by A. C. Duxbury and A. B. Duxbury, 227–28. Dubuque: Wm. C. Brown.

Toy spill, 1992. The first report of the celebrated toy spill appeared in the *Sitka Sentinel.* Punderson, Eben, 1993. "Solved: Mystery of the Wandering Bathtub Toys." *Daily Sitka Sentinel,* September 17, 1993, Sitka Weekend.

A follow-up article asked beachcombers to send letters and maps describing any beached toys they found. Will, S., 1994. "Scientists Trace Odyssey of Bathtub Toys." *Daily Sitka Sentinel.* April 8, 1994.

We analyzed the spill in another article for *EOS.* Ebbesmeyer, C. C., and W. J. Ingraham, Jr., 1994. "Pacific Toy Spill Fuels Ocean Current Pathways Research." *EOS, Transactions of the American Geophysical Union* 75 (37): 425, 427, 430.

And we considered the four types of drifter—natural, determinant, computer, and accidental—together in this recent paper, the first to examine the orbital period of a planetary gyre. Ebbesmeyer, C. C., W. J. Ingraham, Jr., T. C. Royer, and C. E. Grosch, 2007. "Tub Toys Orbit the Pacific Subarctic Gyre." *EOS, Transactions of the American Geophysical Union* 88 (1): 1, 4.

Children's books on the toy spill. Several have been published to enthusiastic receptions, indicating the deep chords struck by the event and what it has revealed about the sea. Bunting, Eve, with illustrations by David Wisniewski, 1997. *Ducky.* Boston: Clarion Books.

Carle, Eric, 2005. *10 Little Rubber Ducks.* New York: HarperCollins.

Burns, Loree Griffin, 2007. *Tracking Trash.* Boston: Houghton Mifflin.

Dabob Bay

Ron Kollmeyer (who followed in Cliff Barnes's footsteps to work for the International Ice Patrol) paved the way for me with his pursuit of water slabs. Kollmeyer, R. C., 1965. "Water Properties and Circulation in Dabob Bay, Autumn 1962." Master of Science thesis, University of Washington.

I recounted my pursuit of the slabs in my own dissertation. Ebbesmeyer, C. C., 1973. "Some Observations of Medium Scale Water Parcels in a Fjord: Dabob Bay, Washington." Ph.D. thesis, University of Washington.

This dissertation developed into an article. Ebbesmeyer, C. C., C. A. Barnes, and C. W. Langley, 1975. "Application of an Advective-Diffusive Equation to a Water Parcel Observed in a Fjord." *Estuarine and Coastal Marine Science* 3: 249–68.

Derelict Vessels

See Vessels.

Drift Cards and Bottles

After a lifetime of chasing drifters, I came to the conclusion that it takes hundreds of thousands of drifters to delineate the processes in a sizeable water body. Self-addressed plastic drift cards became a standard oceanographic instrument in the early 1970s. Hundreds of thousands were released in the 1970s and 1980s, and some still drift in after more than thirty years, returning valuable information on the fate of plastic in the sea and the memory of the gyres. Fifty thousand cards released off South Africa, described in this report, revealed the orbital periods of the Magid, Navigator, and Penguin gyres. Shannon, L. V., G. H. Stander, and J. A. Campbell, 1973. *Oceanic Circulation Deduced from Plastic Drift Cards.* Sea Fisheries Branch Investigation Report No. 108. Cape Town: Republic of South Africa Department of Industries.

More than forty thousand drifters enabled me to see where flotsam accumulates in relation to eddies in the Strait of Juan de Fuca. Sauers, K. A., T. Klinger, C. A. Coomes, and C. C. Ebbesmeyer, 2003. "Synthesis of 41,300 Drift Cards Released in Juan de Fuca Strait (1975–2002)." *2003 Georgia Basin Puget Sound Research Conference Proceedings* 1: 1–12. Olympia, Washington: Puget Sound Action Team.

Eighty-five thousand drifters released in the Gulf of Mexico enabled us to see that flotsam accumulates on the same beaches on Florida's east coast and the southern Texas coast where turtles lay their eggs. Lugo-Fernández, A., M. V. Morin, C. C. Ebbesmeyer, and C. F. Marshall, 2001. "Gulf of Mexico Historic (1955–1987) Surface Drifter Data Analysis." *Journal of Coastal Research* 17 (1): 1–16.

Drifters (general)

Jim Ingraham and I reviewed the range of objects that drift on Northwest waters in an article for the *North Pacific Pilot Charts*, a folio-sized publication carried aboard most vessels. Ebbesmeyer, C. C., and W. J. Ingraham, Jr., 1994. "Some History of Objects Drifting on the Ocean." In *Atlas of Pilot Charts: North Pacific Ocean*, NVPUB108. Washington, D.C.: Defense Mapping Agency, U.S. Department of Defense and U.S. Department of Commerce.

This article contains a wealth of information about all manner of flotsam including message bottles, derelict vessels, and Columbus's clues. Krümmel,

Otto von, 1908. "*Flaschenposten, treibende wracks und andere triftkörper in ihrer bedeutung für die enthullung der meeresströmungen.*" Meereskunde. Heft 7. Berlin: Institut für Meereskunde zu Berlin, Ernst Seigfried Mittler und Sohn.

Several articles in this scientific collection discuss plastic and natural debris worldwide. Coe, J. M., and D. B. Rogers, eds., 1996. *Marine Debris, Sources, Impacts, and Solutions.* Heidelberg and New York: Springer-Verlag.

Schlee summarizes many aspects of oceanography. Schlee, S., 1973. *A History of Oceanography.* London: Robert Hale & Company.

Driftwood

Driftwood piles on the beach contain logs from all over the world. Once in a while I discover successful long-distance drifters—proof that wood can float for years, long enough to cross any ocean. Strong and Skolmen trace driftwood that's orbited the Turtle Gyre. Strong, C. C., and R. G. Skolmen, 1963. "Origin of Drift-Logs on the Beaches of Hawaii." *Nature* 197 (March 2, 1963): 890.

Smith, Rudall, and Keage do the same for wood that's orbited the Penguin Gyre. Smith, J. M. B., P. Rudall, and P. L. Keage, 1989. "Driftwood on Heard Island." *Polar Record* 25 (154): 223–28.

Nansen discusses wood that's transited the Arctic Ocean. Nansen, F., 1911. *In Northern Mists.* Vol. 2. New York: Frederick A. Stokes Company.

Laeyendecker examines wood that's drifted from the Yukon River to Frobisher Bay. Laeyendecker, Dosia, 1993. "Wood and Charcoal Remains from Kodlunarn Island." In *Archeology of the Frobisher Expeditions.* Edited by W. W. Fitzhugh and J. S. Olin, 1993 Washington, D.C.: Smithsonian Institution Press, 164.

Laeyendecker refers to this analysis. Eggertsson, O., 1991. "Driftwood in the Arctic, a Dendrochronological Study." In *Lundqua Reports.* Lund, Sweden: Department of Quaternary Geology, University of Lund.

Ferry Wakes

While studying contaminated sediments along Seattle's waterfront, I noticed that fine-grain sediments were missing beneath the ferry terminal and offshore, and currents were high in the area. I fought to get the size of the sediment grains and the currents near the bottom measured. My student Mike Francisco did the spadework to prove that ferry wakes did indeed affect sediments and currents. Michelsen, T. C., C. D. Boatman, D. Norton, C. C. Ebbesmeyer, T. Floyd, and M. D. Francisco, 1998. "Resuspension and Transport of Contaminated Sediments Along the Seattle Waterfront,

Part 1: Field Investigations and Conceptual Model." *Journal of Marine Engineering* 5: 35–65.

Francisco, M. D., C. C. Ebbesmeyer, C. D. Boatman, and T. C. Michelsen, 1998. "Resuspension and Transport of Contaminated Sediments Along the Seattle Waterfront, Part 2: Resuspension and Transport Mechanisms." *Journal of Marine Engineering* 5: 67–84.

Floating Islands

Don't overlook the twenty-four photographs at the back of this book-length list of annotated references on the elusive isles. Van Duzer, Chet, 2004. *Floating Islands, a Global Bibliography.* Los Altos: Cantor Press.

Garbage Patches

My first encounter with a garbage patch occurred in the Strait of Juan de Fuca. Page 44 of this official study shows the patch of drift sheets collected there. Ebbesmeyer, C. C., J. M. Cox, J. M. Helseth, L. R. Hinchey, and D. W. Thomson, 1979. "Dynamics of Port Angeles Harbor and Approaches, Washington." U.S. Department of Commerce, EPA–600/7–79–252.

Prince Albert I was the first to describe a garbage patch, the one beneath the Azores high-pressure cell in the Columbus Gyre, based on an analysis of 227 bottles released in the North Atlantic. H.S.H. Albert, Prince of Monaco, 1892. "A New Chart of the Currents of the North Atlantic." *Scottish Geographical Magazine* (October 1892): 528–31.

Dean Bumpus directed the release of some 150,000 bottles along the United States' east coast, many of which wound up in the Azores patch. This important paper includes an analysis of 480 transatlantic bottles Bump compiled in the archives of the Woods Hole Oceanographic Institution. Scheltema, Rudolf S., 1966. "Evidence for Trans-Atlantic Transport of Gastropod Larvae Belonging to the Genus *Cymatium.*" *Deep-Sea Research* 13: 83–95.

See footnote 12, "Bottle-drift on the Azores," on pages 460–62 of this classic study. Guppy, H. B., 1917. *Plants, Seeds, and Currents in the West Indies and Azores: The Results of Investigations Carried out in those Regions Between 1906 and 1914.* London: Williams and Norgate.

Glass Balls

Amos Wood's groundbreaking volume is the bible for students and aficionados of glass fishing floats. Wood, Amos, 1967. *Beachcombing for Japanese Glass Floats.* Hillsboro, Oregon: Binford & Mort.

Alan Rammer and Walt Pich have continued Wood's work. Rammer has published two books on the pricing of glass balls and Pich has published this and one other on beachcombing. Pich, Walter, 1997. *Beachcomber's Guide to the Northwest.* Ocean Shores, Washington: Walter C. Pich Publishing.

Global Warming and Ocean Circulation

Bryden, Harry L., Hannah R. Longworth, and Stuart A. Cunningham, 2005. "Slowing of the Atlantic Meridional Overturning Circulation at 25° N." *Nature* 438 (December 1, 2005): 655–57.

Halobates (oceanic water striders)

"Halobates" derives from the Latin *halo* (salty) and *bates* (walker). These oceanic water striders are masters of the two-dimensional universe known as the Floating World. Over the years, I noticed that Charlie Moore trawled up many tiny black halobates along with much finely ground plastic. Gliding among all that plastic must greatly alter their environment and life history. Charlie has counted plankton with respect to plastic, but I do not know how plastic counts with respect to halobates.

Lanna Cheng offers a good overview of these remarkable creatures. Cheng, Lanna, 1972. "Skaters of the Seas." *Oceans* 5: 6, 54–55.

She has also written a technical overview of both coastal halobates (thirty-nine species) and pelagic (five species). Cheng, Lanna, 1985. "Biology of Halobates (Heteroptera: Gerridae)." *Annual Review of Entomology* 30: 111–35.

This article provides a historical introduction to halobates. McHugh, J. L., and Hilary B. Moore, 1956. "Treaders of the Sea." *Bulletin of the International Oceanographic Foundation* 2 (July 2, 1956): 104–7.

Scheltema discusses halobates' tiny size. Scheltema, R. S., 1978. "Ocean Insects." *Oceanus* 14 (3): 8–12. See also Milne, L. J., and M. Milne, 1978. "Insects of the Water Surface." *Scientific American* 238 (4): 134–42.

For stimulating writing (with little math) on the physics of water striders, see Vogel, Steven, 1994. *Life in Moving Fluids: The Physical Biology of Flow,* 2nd ed. Princeton, New Jersey: Princeton University Press.

Henry the Navigator, Prince

Though no explorer himself, Portugal's Prince Henry greatly advanced the cause of exploration when he founded the first school of oceanography. Guill,

James H., 1980. "Vila do Infante (Prince-Town), the First School of Oceanography in the Modern Era: An Essay." In *Oceanography of the Past: Proceedings of the Third International Congress on the History of Oceanography*, edited by M. Sears and D. Merriman. Woods Hole, Massachusetts: Woods Hole Oceanographic Institution, 596–605.

Human Remains Afloat

Bill Haglund and I reported our forensic findings in two papers. Ebbesmeyer, C. C., W. P. Haglund, 1993. "Drift Trajectories of a Floating Human Body Simulated in a Hydraulic Model of Puget Sound." *Journal of Forensic Sciences* 39 (1): 231–40.

Ebbesmeyer, C. C., and W. P. Haglund, 2002. "Floating Remains on Pacific Northwest Waters." In *Advances in Forensic Taphonomy: Method, Theory, and Archaeological Perspectives*, edited by W. P. Haglund and M. H. Sorg. Boca Raton: CRC Press, 219–40.

Specific gravity of human bodies. Whether a body floats depends on many factors, including clothing, lung capacity, gaseous decomposition, and water density (which reflects temperature and salinity). Despite the obvious practical importance of this question, only one study to my knowledge has investigated it, and only in a narrow range of conditions. It estimates that 69 percent of humans would float in seawater. My working hypothesis, based on my own experience, is that if many studies were to be performed under a wide range of conditions, they would find that about half of bodies sink and half float. Donoghue, E. R., and S. C. Minnigerode, 1977. "Human Body Buoyancy: A Study of 98 Men." *Journal of Forensic Science* 22 (3): 573–79.

Durability of human bodies in seawater. This study describes bones that have survived for a thousand years. Arnaud, G., S. Arnaud, A. Ascenzi, E. Bonucci, and G. Graziani, 1978. "On the Problem of the Preservation of Human Bone in Sea-Water." *Journal of Human Evolution* 7: 409–20.

Bodies have floated for four months in the North Sea. Giertsen, Johan Christopher, and Inge Morild, 1989. "Seafaring Bodies." *American Journal of Forensic Medicine and Pathology* 10 (1): 25–27.

Flesh has survived for three years in Puget Sound. See case 11 here. Haglund, William D., 1993. "Disappearance of Soft Tissue and the Disarticulation of Human Remains from Aqueous Environments." *Journal of Forensic Sciences* 38 (4): 806–15.

Icebergs and Ice Islands

See also Melville, George W.

Tom Budinger was Cliff Barnes's premier student, a polymath who was enrolled simultaneously in the UW schools of law, medicine, and oceanography. The authorities declared this unacceptable, so he chose oceanography. Cliff told me Tom had been in line to be an astronaut. Tom recounted Cliff's story of the giant iceberg and the World War II convoy in this encyclopedia article. Budinger, T. F., 1974. "Icebergs and Pack Ice." In *Encyclopedia Britannica*, 15th ed., 9:154–61.

For an overview of ice islands, see Jeffries, M. O., 1992. "Arctic Shelves and Ice Islands: Origin, Growth and Disintegration, Physical Characteristics, Structural-Stratigraphic Variability, and Dynamics." *Reviews of Geophysics* 30 (3): 245–67.

Storkerson's ride was front-page news. "Five on an Ice Cake Test Polar Current." *New York Times*, February 26, 1919.

It followed the equally dramatic *Polaris* expedition. Blake, E. V., 1874. *Arctic Experiences: Containing Captain George E. Tyson's Wonderful Drift on the Ice Floe, a History of the* Polaris *Expedition, the Cruise of the Tigress, etc.* New York: Harper Brothers.

Subsequent expeditions have further probed the movements of Arctic ice. Smith, E. H., 1931. *The Marion Expedition to Davis Strait and Baffin Bay, 1928, Scientific Results, Part 3: Arctic Ice, with Especial Reference to Its Distribution to the North Atlantic Ocean.* Coast Guard Bulletin No. 19. Washington: U.S. Treasury Department.

Iceland

I first learned of drifting objects used by the Vikings from a dissertation by an Icelandic student who fortunately sent Cliff Barnes a copy. Unnsteinn Stefánsson carefully located the mentions of drifting objects in the sagas. I correlated these with modern maps of Iceland. Stefánsson, Unnsteinn, 1962. *North Icelandic Water (Atvinnudeild Haskolans—Fiskideild).* Ph.D. dissertation, Department of Fisheries, University Research Institute, Reykjavik.

Palsson, H., and P. Edwards, trans., 1972. *The Book of Settlements (Landnamabok).* Winnipeg: University of Manitoba Icelandic Studies.

Jones, G., trans., 1960. *Egil's Saga.* Syracuse, New York: Syracuse University Press.

Iron, Drift

The prominent mining engineer Thomas Arthur Rickard published three seminal works on iron afloat in drifting wood and its impact on native cultures. Rickard, T. A., 1932. "The Knowledge and Use of Iron Among the South Sea Islanders." *Journal of the Royal Anthropological Institute* 62: 1–22.

Rickard, T. A., 1934. "Drift Iron, a Fortuitous Factor in Primitive Culture." *Geographical Review* 24 (October 1934): 4.

Rickard, T. A., 1938. "The Use of Iron and Copper by the Indians of British Columbia." *British Columbia Historical Quarterly* 3 (1): 25–50.

One of my heroes is Juan Francisco de la Bodega y Quadra, Spain's answer to Captain Cook, who had to watch as his men were slaughtered for the iron in their boat. I learned of this incident from a then-unpublished 1993 manuscript, "Incident at la Punta de las Matires," which the since-deceased H. C. Taylor shared with me. It had been submitted to *Papers in Honor of Keith Murray*, but I do not know if it has ever been published.

Japanese Drifters

Plummer, Katherine, 1991. *The Shogun's Reluctant Ambassadors: Japanese Sea Drifters in the North Pacific.* 3rd ed., revised. Portland: Oregon Historical Society Press.

MacDonald, Ranald

Lewis, W. S., and N. Murakami, 1990. *Ranald MacDonald, the Narrative of his Life, 1824–1894.* Portland: Oregon Historical Society Press.

Roe, J. A., 1997. *Ranald MacDonald, Pacific Rim Adventurer.* Pullman, Washington: Washington State University Press.

Maury, Matthew Fontaine

Maury's descendant Anne Fontaine Maury cites a memorandum among the papers of her father C. W. Maury. "Amerigo, Ammerigo, Merigo . . . Amerigo is an Italianized form of an old German word, which in medieval French became Amaury which means 'the steadfast.'" Matthew Fontaine Maury's steadfast belief in the Bible's marine pathways led to the founding of oceanography. His steadfast love of his Virginia home led him to quit his naval post and join the Confederacy. His amazing coordination of ship reports showed another sort of steadfastness. See their family history. Anne Fontaine Maury,

1941. *Intimate Virginiana, a Century of Maury Travels by Land and Sea.* Richmond: Dietz Press, 311–34.

This broader history describes Maury's role in developing floating mines. Perry, M. F., 1965. *Infernal Machines: The Story of Confederate Submarine and Mine Warfare.* Baton Rouge: Louisiana State University Press, 3.

Melville, George W.

Melville (and others) recounted his harrowing adventures on the *Jeannette* voyage, which left him shipwrecked on an ice floe and inspired him to launch unmanned drifters on the Arctic ice. Melville, G. W., 1897. "The Drift of the *Jeannette.*" *Proceedings of the American Philosophical Society* 36: 156.

Guttridge, L. F., 1986. *Icebound: The* Jeannette *Expedition's Quest for the North Pole.* Annapolis: Naval Institute Press.

Melville also explained his drift-cask design. Melville, G. W., 1898. "A Proposed System of Drift Casks to Determine the Direction of the Circumpolar Currents." *Bulletin of the Geographical Society of Philadelphia* 2 (3): 41–45.

For where the drift casks washed up, see Smith, E. H., 1931. "The Marion Expedition to Davis Strait and Baffin Bay, Scientific Results, Part 3: Arctic Ice, with Especial Reference to its Distribution to the North Atlantic Ocean." U.S. Coast Guard Bulletin 19: 26–29.

Messages in Bottles

The first known publication of Edgar Allan Poe's "MS. Found in a Bottle" was in the *Baltimore Saturday Visitor,* October 1833. Poe himself indicated that it was first published in 1831.

The data used in my timeline of early bottle papers are contained in three reports that I believe were compiled by Commander Alexander Becher. The first is accompanied by a chart dated February 1, 1843, and titled "Bottle Chart of the Atlantic Ocean by A. B. Becher, Commander Royal Navy." In *Nautical Magazine and Naval Chronicle* 12, no. 2 (February 1843): 181–84.

The second, from November 1852, is uncredited. "The Bottle Chart of the Atlantic Ocean." *Nautical Magazine and Naval Chronicle* 21 (11): 4D, 569–72.

The third, from December 1852, also uncredited, includes a table of bottle papers. *Nautical Magazine and Naval Chronicle* 21 (12): 671–72.

One of the largest systematic studies conducted off any coast consisted of at least 148,000 bottles released over seventeen years by these California researchers. Crowe, F. J., and R. A. Schwartzlose, 1972. "Release and Recovery Records of Drift Bottles in the California Current Region, 1955 through 1971." *Califor-*

nia Cooperative Oceanic Fisheries Investigations, Atlas No. 16. Data Collection and Processing Group, Marine Life Research Program. La Jolla, California: Scripps Institution of Oceanography.

Mines

Ten North Pacific pilot charts showing mines drifting across the Turtle Gyre, from: U.S. Hydrographic Office, Pilot charts of the North Pacific Ocean, No. 1401, issued monthly, Washington, D.C. The first six charts have been located; the last four remain missing. Readers are requested to notify Curt Ebbesmeyer of the whereabouts of the missing charts.

Chart issue date	*Dates of mine sightings shown on chart*
June 1946+	13 February- 28 March 1946
July 1946+	29 March- 28 April 1946
August 1946+	29 April- 28 May 1946
September 1946+	29 May- 30 June 1946
October 1946+	1 July- 31 July 1946
November 1946+	1 August- 31 August 1946
December 1946*	1 September- 30 September 1946
January 1947*	1 October- 31 October 1946
February 1947*	1 November- 30 November 1946
March 1947*	1 December- 31 December 1946

* Chart not located.

+ If not for the detective work of Pam Mofjeld, librarian at the Fisheries and Oceanography Library, University of Washington, and Paul Leverenz, Scripps Institution of Oceanography, these charts might not have been located and reproduced. Given the large numbers of mines adrift after World War II, I thought it would be a simple task to locate information on sightings at sea and reports from bomb disposal squads who had deactivated the mines that had drifted onto the beaches. This was not to be the case. The North Pacific Pilot charts, issued monthly to ships by the Defense Mapping Agency, showed hundreds of mines. Unfortunately, the Defense Mapping Agency had not marked the pilot charts for permanent library archival, but rather for disposal after a few months. Luckily, librarians at Scripps Institution of Oceanography, La Jolla, California, saved copies of six of the charts.

I was never able to trace the authorship or origin of the detailed map showing where and when submarine mines washed ashore along Oregon and Washington that Cliff Barnes left secreted in a steel army box. But Taivo Laevastu left this fine work to Jim Ingraham who, knowing of my deep interest in mines, gave it to me. Johnson, E.A., and D.A. Katcher, 1947. *Mines Against Japan.*

Released for public distribution June 1973. White Oak, Maryland: Silver Spring: Naval Ordnance Laboratory.

This volume treats the same subject. Hartmann, G. K., and S. C. Truver, 1991. *Weapons That Wait: Mine Warfare in the U.S. Navy.* Annapolis: Naval Institute Press.

The estimate of thirty-five thousand mines adrift in the North Pacific was published three years after the war ended. Bristol, J. A., 1948. "Here Come the Jap Mines." *Saturday Evening Post* 220 (March 20, 1948): 12.

Mines washed up in Hawaii in 1955 and 1956. Lott, A. S., 1959. *Most Dangerous Sea: A History of Mine Warfare, and an Account of U.S. Navy Mine Warfare Operations in World War II and Korea.* Annapolis: U.S. Naval Institute, 264.

Prince Albert I of Monaco published three papers on mines adrift after World War I, in 1918, 1919, and 1920 in *Comptes rendus hebdomadaires des seances de l'Academie des sciences* 167: 1049–56; 169: 562–66; and 170: 778–82.

The first paper was translated and republished. H.S.H. Albert, Prince of Monaco, 1919. "Floating Mines in the North Atlantic and Arctic Oceans." *Scientific American* 120, no. 16 (April 19, 1919): 394–95, 416.

Albert also discussed mines in an address to the National Academy of Sciences on April 25, 1921, subsequently published. H.S.H. Albert, Prince of Monaco, 1921. "Studies of the Ocean." *Scientific Monthly* 13, no. 2 (August 1921): 171–85.

Davis summarizes Prince Albert's thoughts on mines. Davis, W. M., 1920. "The Drift of Mines in the North Atlantic." *Geographical Record* 10 (December 1920): 419–20.

Perry describes ship sinkings by Confederate mines. Perry, M. F., 1965. *Infernal Machines: The Story of Confederate Submarine and Mine Warfare.* Baton Rouge: Louisiana State University Press, 199.

Nike Sneakers

See Container Spills.

Okubo, Akira

Our collaboration. Akira and I published six peer-reviewed papers together, beginning with our study of Seattle sewage. We executed this paper from a conventional Eulerian viewpoint for the sake of speed, in time with our Metro contract. But the Lagrangian coordinate system is a more natural reference frame for studying drifters, though it's almost entirely unfamiliar to those who study them at sea. Okubo, A., and C. C. Ebbesmeyer, 1976. "Determination of Vorticity, Divergence, and Deformation Rates from Analysis of Drogue Observations." *Deep-Sea Research* 23: 349–52.

When we had time, we recast that paper into Lagrangian terms. Okubo, A., C. C. Ebbesmeyer, and J. M. Helseth, 1976. "Determination of Lagrangian Deformations from Analysis of Current Followers." *Journal of Physical Oceanography* 6, no. 4 (July 1976): 524–27.

Next we tackled mosquito swarming. Okubo, A., C. Chiang, and C. C. Ebbesmeyer, 1977. "Acceleration Field of Individual Midges, *Anarete Pritchardi* Kim Within A Swarm." *Journal of Canadian Entomology* 109: 149–56.

This led to a paper examining the migration of icebergs. Ebbesmeyer, C. C., A. Okubo, and J. M. Helseth, 1980. "Description of Iceberg Probability Between Baffin Bay and the Grand Banks Using a Stochastic Model." *Deep-Sea Research* 27A: 975–86.

Okubo, A., C. C. Ebbesmeyer, and B. G. Sanderson, 1983. "Lagrangian Diffusion Equation and its Application to Oceanic Dispersion." *Journal of the Oceanographical Society of Japan* 39, no. 5 (October 1983): 259–66.

Richard Strickland's discovery of a message bottle led to perhaps the first scientific paper to result from a kayaker stopping for a call of nature. Ebbesmeyer, C. C., W. J. Ingraham, R. McKinnon, A. Okubo, R. Strickland, D. P. Wang, and P. Willing, 1993. "Bottle Appeal for the Release of China's Dissident Wei Jingsheng Drifts Across the Pacific." *EOS, Transactions of the American Geophysical Union* 74 (16): 193–94.

Akira's ashes. *See* Ashes.

Osiris

The book that inspired my father throughout his life also led me to the oceanographic legend of Osiris. Pike, A., 1871. *Morals and Dogma of the Ancient and Accepted Scottish Rite of Freemasonry.* Reprinted 1930. Richmond, Virginia: L. H. Jenkins.

Pacific, Steamship

I spent months combing microfilms of Pacific Northwest newspapers for accounts of debris washing ashore from the wreck of the SS *Pacific*, then assembled all the sightings on a map and labeled the wash-ups by date. At UC-Berkeley's Bancroft Library I found a letter from a firsthand observer describing a line of debris extending all the way up the Strait of Juan de Fuca. This remains the only data set for debris washed from the coast into inland waters by a storm. Ebbesmeyer, C. C., J. M. Cox, and B. L. Salem, 1991. "1875 floatable wreckage driven inland through the Strait of Juan de Fuca." *Puget Sound Research '91 Proceedings* 1: 75–85. Olympia: Puget Sound Water Quality Authority.

Pangborn, Clyde

There's no space for all the harrowing incidents that occurred during Pangborn's forty-one-hour transpacific flight. However my favorite is the one in which, at fourteen thousand feet in the air, Pang climbed out onto the wing struts to hammer loose the landing gear, first one side, then the other. Seek out the rare book by Pang's close friend Carl Cleveland, whom I interviewed. Cleveland, Carl M., 1978. *Upside-Down Pangborn, King of the Barnstormers.* Glendale, California: Aviation Book Company.

Ripley's Believe It Or Not! cartoon concerning the drift of *Miss Veedol's* wheel appeared in the *Seattle Post-Intelligencer,* July 5, 1933. The explanatory footnote for the July 6 *Ripley's* cartoon provided additional details concerning "the homesick wheel"—how halibut fisherman Ivar Baggen of Seattle found it floating forty miles southwest of Cape Flattery, as reported in the *Seattle Post-Intelligencer,* February 27, 1933, HH 3.

Pioneer Flotsametricians

Gumprecht, T. E., 1854. *Zeitschrift fur allgemeine erdkunde,* vol. 3. Berlin: Reimer.

Alas, this article contains no references. Carruthers, J. N., 1956. "'Bottle Post' and Other Drifts." *Journal of the Institute of Navigation* 9 (3): 261–81.

Plastic, Invention of

For the growth in piano sales, spurring ivory consumption and depletion, see Ehrlich, Cyril, 1976. *The Piano: A History.* London: J. M. Dent & Sons, 128–31.

This marketing booklet, circa 1892, includes extracts from contemporary newspaper articles on the ivory shortage that spurred the search for a substitute material. *Eburnea: The Synthetic Ivory.* Lancaster, England: Eburnea Coy's Works.

For another revealing catalog, circa 1875, see *Ivory.* Hamburg, Germany: Hein. Ad. Meyer.

John Wesley Hyatt recounted how he created celluloid, with remarks on exploding billiard balls and other problems with nitrocellulose, when he received the Society of Chemical Industry's Perkins Medal. Hyatt, John Wesley, 1914. "Address of Acceptance." *Journal of Industrial and Engineering Chemistry* 6, no. 2 (February 1914): 158–61. In the same issue, see also remarks by Leo H. Baekeland, 90–91, and Charles F. Chandler,156–58.

Meikle, Jeffrey L., 1995. *American Plastic: A Cultural History.* New Brunswick, New Jersey: Rutgers University Press.

Plastic Pollution

Nurdles. Environmental Protection Agency, 1992. "Plastic Pellets in the Aquatic Environment, Sources and Recommendations." Final Report. EPA 842-B–92–010.

Two popular books brought plastics' hormonal and reproductive effects to the public's attention. Colborn, Theo, Dianne Dumanoski, and Jonathan Peters, 1996. *Our Stolen Future.* New York: Dutton.

Cadbury, Deborah, 1999. *Altering Eden: The Feminization of Nature.* New York: St. Martin's Press.

See also, Haeba, Maher H., Klára Hilscherová, Edita Mazurová, and Ludek Bláha, 2008. "Selected Endocrine Disrupting Compounds (Vinclozolin, Flutamide, Ketoconazole and Dicofol): Effects on Survival, Occurrence of Males, Growth, Molting and Reproduction of Daphnia Magna." *Environmental Science and Pollution Research International* 50, no. 3 (May 2008): 222–27.

Oregon Department of Human Services Environmental Toxicology. "About PBDE Flame Retardants." Fact sheet, undated.

Swan, S. H., E. P. Elkin, and L. Fenster, 1997. "Have Sperm Densities Declined? A Reanalysis of Global Trend Data." *Environmental Health Perspectives* 105 (11): 1228–32.

Adsorption of toxic chemicals. Mato, Y., T. Isobe, H. Takada, H. Kanehiro, C. Ohtake, and T. Kaminuma, 2001. "Plastic Resin Pellets as a Transport Medium for Toxic Chemicals in the Marine Environment." *Environmental Science Technology* 35, no. 2 (January 15, 2001): 318–24.

Raloff, Janet, 2001. "Plastic Debris Picks up Ocean Toxics." *Science News* 159, no. 5 (February 3, 2001): 79.

In the Great North Pacific Garbage Patch. Moore, C. J., S. L. Moore, M. K. Leecaster, and S. B. Weisberg, 2001. "A Comparison of Plastic and Plankton in the North Pacific Central Gyre." *Marine Pollution Bulletin* 42: 1297–300.

Prince Albert I

See Mines.

Pumice

The origin of life. Ebbesmeyer, C. C., and W. J. Ingraham, Jr., 1999. "Pumice and Mines Afloat on the Sea." In "Tribute to Akira Okubo." *Oceanography* 12 (1): 17–21.

Vikings. For floating pumice from the 1362 eruption of Vatna volcano, see Byock, Jesse L., 2001. *Viking Age Iceland.* London/New York: Penguin Books, 61–62.

From Krakatoa. For the definitive account, see Simkin, T., and R. S. Fiske, 1983. *Krakatau 1883: The Volcanic Eruption and Its Effects.* Washington, D.C.: Smithsonian Institution Press.

Frick and Kent found pumice from the 1983 Krakatoa eruption only on the Indian Ocean, not on Atlantic beaches. Frick, C., and L. E. Kent, 1984. "Drift pumice in the Indian and South Atlantic Oceans." *Transactions of the Geological Society of South Africa* 87 (1): 19–33.

Transpacific. Pumice crossed the Pacific, from Mexico to the Philippines. Richards, A. F., 1958. "Transpacific Distribution of Floating Pumice from Isla San Benedicto, Mexico." *Deep-Sea Research* 5: 29–35.

South Sandwich Islands. Rafts of pumice from the 1962 South Sandwich eruption reached Australia and New Zealand. Wace, Nigel, 1991. "Garbage in the Oceans." *Bogong, Journal of the Canberra and Southeast Regional Environment Centre* 12, no. 1 (Autumn 1991): 15–18.

This study provides an overview of the pumice drift. Coombs, D. S., and C. A. Landis, 1966. "Pumice from the South Sandwich Eruption of March 1962 Reaches New Zealand." *Nature* 209 (5020): 289–90.

Gass and colleagues map the edge of the two-thousand-square-mile pumice raft. Gass, I. G., P. G. Harris, and M. W. Holdgate, 1963. "Pumice Eruption in the Area of the South Sandwich Islands." *Geological Magazine* 100, no. 4 (July–August 1963): 321–30.

Jokiel and Cox track the global drift of pumice from both Krakatoa and the South Sandwich Islands. Jokiel, P. L., and E. F. Cox, 2003. "Drift Pumice at Christmas Island and Hawaii: Evidence of Oceanic Dispersal Patterns." *Marine Geology* 202: 121–33.

Sea Beans

There are three "beaner" bibles. Gunn, C. R., and J. V. Dennis, 1976. *World Guide to Tropical Drift Seeds and Fruits.* Originally a Demeter Press Book from Times Books. Reprinted 1999, Malabar, Florida: Krieger Publishing.

Perry IV, E., and J. V. Dennis, 2003. *Sea-Beans from the Tropics.* Malabar, Florida: Krieger Publishing.

Nelson, E. C., 2000. *Sea Beans and Nickar Nuts.* Handbook No. 10. London: Botanical Society of the British Isles.

See also Armstrong, Wayne P., 1990. "Seed Voyagers." *Pacific Discovery* (Summer 1990): 32–40.

This newsletter is indispensable. Perry, Ed, ed. *The Drifting Seed.* Melbourne, Florida,

Sea Glass

Lambert, C. S., and P. Hanbery, 2001. *Sea Glass Chronicles.* Rockport, Maine: Down East Books.

Sewage Microlayer

In this paper, my colleagues and I showed that despite being treated to a high-quality secondary level, oil and grease composing 10 percent of sewage effluent rose several hundred feet from the discharge pipe to float on the sea surface. Nevertheless, environmental protocols persist in dictating that measurements be taken one meter below the surface—below these floating oils and greases. And it is in these greases that contaminants—heavy metals, carcinogenic hydrocarbons—reside. Word, J. Q., C. D. Boatman, C. C. Ebbesmeyer, R. E. Finger, S. Fischnaller, and Q. J. Stober, 1990. "Vertical Transport of Effluent Material to the Surface of Marine Waters." *Oceanic Processes in Marine Pollution* 6: 134–49.

This book examines the physics and chemistry of the surface layer where sewage collects. Monahan, E. C., and M. A. Van Patten, 1989. *The Climate and Health Implications of Bubble-Mediated Sea-Air Exchange.* Connecticut Sea Grant College Program CT-SG–89–06.

Shoreline Accumulation

For thirty years I used drift cards to study the water bodies of the Puget Sound system. I soon learned that the currents deposited the cards on certain shorelines. The gyres in this early study are eddies formed by swirling tidal currents, analogous to the immense gyres swirling between the continents. Ebbesmeyer, C. C., C. A. Coomes, J. M. Cox, and B. L. Salem, 1991. "Eddy Induced Beaching of Floatable Materials in the Eastern Strait of Juan de Fuca." In *Puget Sound Research '91 Proceedings.* Olympia: Puget Sound Water Quality Authority, 86–98.

Sauers, K. A., T. Klinger, C. A. Coomes, and C. C. Ebbesmeyer, 2004. "Synthesis of 41,300 drift cards released in Juan de Fuca Strait (1975–2002)." In T. W. Droscher and D. A. Fraser, eds., Proceedings of the 2003 Georgia Basin/Puget Sound Research Conference. Puget Sound Action Team, Olympia, Washington. Available at http://www.psat.wa.gov/Publications/03_proceedings/start.htm.

Snarks

See Water Slabs.

Theophrastus

In a personal communication dated December 20, 1991, the Theophrastus scholar R. W. Sharples of University College, London, informed me of Peter Steinmetz's assertion regarding Aristotle's great student. "Theophrastus is the first physicist to have examined the phenomena of currents on a broad basis in the most various ways." What Steinmetz means by "currents" is open to debate but most likely he refers to air currents. Steinmetz, Peter, 1964. *Die Physik des Theophrastos von Eresos* (Palingenesia, I) p. 328. Bad Homburg: Max Gehlen.

Tub Toys

See Container Spills.

Vessels, Drifting and Derelict

Unfortunately, there's no comprehensive global treatment of the countless derelict vessels that have drifted across the Floating World. But these studies examine derelicts in the two most familiar oceanic regions.

North Atlantic. This article by Richardson includes a statistical summary of derelicts adrift during the last days of sail in the North Atlantic. Note especially section 4, concerning twenty-seven thousand tree trunks lost there. Richardson, P. L., 1985. "Drifting Derelicts in the North Atlantic 1883–1902." *Progress in Oceanography* 14: 463–83.

Richardson, P. L., 1985. "Derelicts and Drifters, Old Abandoned Sailing Ships and New Satellite-Tracked Buoys Tell Us Where Some Ocean Currents Come From and Where They're Going." *Natural History* 94 (6): 42–49.

North Pacific. This is the seminal survey of Japanese vessels drifting across the North Pacific. Brooks, C. W., 1875. "Japanese Wrecks Stranded and Picked Up Adrift in the North Pacific Ocean." *Proceedings of the California Academy of Sciences* 6: 50–66.

Davis, H., 1872. *Record of Japanese Vessels Driven Upon the North-West Coast of America and its Outlying Islands.* Worcester: American Antiquarian Society.

Meggers, Betty J., and Clifford Evans, 1966. "A Transpacific Contact in 3000 B.C." *Scientific American* 214 (1): 28–35.

Plummer, Katherine, 1991. *The Shogun's Reluctant Ambassadors, Japanese Sea Drifters in the North Pacific.* 3rd ed., rev. Portland: Oregon Historical Society Press.

Water Slabs (snarks)

Puget Sound. Ebbesmeyer, C. C., C. A. Barnes, and C. W. Langley, 1975. "Application of an Advective-Diffusive Equation to a Water Parcel Observed in a Fjord." *Estuarine and Coastal Marine Science* 3: 249–68.

North Atlantic. This is one of six papers in a volume I jointly authored describing the results of the POLYMODE Local Dynamics Experiment. Two recount how two slabs were revisited several times during two months at sea—an extended period as such research goes—but a short spell in the long lives of these slabs. Here we examined ten snarks from all over the North Atlantic found just south of Bermuda. Ebbesmeyer, C. C., B. A. Taft, J. C. McWilliams, C. Y. Shen, S. C. Riser, H. T. Rossby, P. E. Biscaye, and H. G. Östlund, 1986. "Detection, Structure, and Origin of Extreme Anomalies in a Western Atlantic Oceanographic Section." *Journal of Physical Oceanography* 16, no. 3, (March 1986): 591–612.

Gulf Stream. Glenn, S. M., and C. C. Ebbesmeyer, 1994. "Observations of Gulf Stream Frontal Eddies in the Vicinity of Cape Hatteras." *Journal of Geophysical Research* 99 (C3): 5047–56.

Glenn, S. M., and C. C. Ebbesmeyer, 1994. "The Structure and Propagation of a Gulf Stream Frontal Eddy Along the North Carolina Shelf Break." *Journal of Geophysical Research* 99 (C3): 5029–46.

Whaling

Aleuts in skin baidarkas hunted whales with poisoned lances. Heizer, R. F, 1938. "Aconite Poison Whaling in Asia and America: An Aleutian Transfer to the New World." *Bulletin of the Bureau of American Ethnology* 133: 417–68.

Heizer, Robert F., 1938. "Aconite Arrow Poison in the Old and New World." *Journal of the Washington Academy of Sciences* 28 (8): 358–64.

Yasuyori's Stupa

McCullough, H. C., trans., 1988. *The Tale of the Heike.* Stanford, California: Stanford University Press.

Index

Note: Page numbers in *italics* indicate photographs and illustrations.